T0332703

SEQUENTIAL LOGIC SYNTHESIS

THE KLUWER INTERNATIONAL SERIES
IN ENGINEERING AND COMPUTER SCIENCE

VLSI, COMPUTER ARCHITECTURE AND
DIGITAL SIGNAL PROCESSING
Consulting Editor
Jonathan Allen

Latest Titles

SEQUENTIAL LOGIC SYNTHESIS

by

Pranav Ashar
University of California/Berkeley

Srinivas Devadas
Massachusetts Institute of Technology

A. Richard Newton
University of California/Berkeley

Kluwer Academic Publishers
Boston/Dordrecht/London

Distributors for North America:
Kluwer Academic Publishers
101 Philip Drive
Assinippi Park
Norwell, Massachusetts 02061 USA

Distributors for all other countries:
Kluwer Academic Publishers Group
Distribution Centre
Post Office Box 322
3300 AH Dordrecht, THE NETHERLANDS

Library of Congress Cataloging-in-Publication Data

Ashar, Pranav.
  Sequential logic synthesis / by Pranav Ashar, Srinivas Devadas, A.
Richard Newton.
      p. cm. -- (The Kluwer International series in engineering and
computer science)
  Includes bibliographical references and index.
    ISBN 0-7923-9187-X (alk. paper)
      1. Logic circuits--Design and construction--Data processing.
  2. Logic design--Data processing. 3. Computer-aided design.
  4. Integrated circuits--Very large scale integration--Design and
construction--Data processing.   I. Devadas, Srinivas. II. Newton,
A. Richard (Arthur Richard), 1951- . III. Title. IV. Series
TK7868.L6A84 1991
621.39'5--dc20   91-37874
                                                                      CIP

Printed on acid-free paper.

Printed in the United States of America

Contents

List of Figures

List of Tables

Preface

Automata theory forms a cornerstone of digital Very Large Scale Integrated (VLSI) system design. This book deals exclusively with finite automata theory and practice. The extensive use of finite state automata, finite state machines (FSMs) or simply sequential logic in VLSI circuits and the recent proliferation of Computer-Aided Design (CAD) research in the area of FSM synthesis has prompted the writing of this book.

Historically, finite automata were first used to model neuron nets by McCulloch and Pitts [69] in 1943. Kleene [52] considered regular expressions and modeled the neuron nets of McCulloch and Pitts by finite automata, proving the equivalence of the two concepts. Similar models were considered by Huffman [48], Moore [75], and Mealy [70] in the '50s.

The early sixties saw the increasing use of finite automata in switching circuit or switching relay design. The problem of finding a minimal realization of a FSM became increasingly important. Considerable theoretical work in the algebraic structure of FSMs was carried out, notably by Armstrong [1], Hartmanis [41], and others. Applications of the developed theory led to algorithms that minimize the number of states in a FSM, encode the states of a FSM and decompose a FSM into smaller submachines. The work of this era has been comprehensively presented in several books − for instance those by Hartmanis and Stearns [44], Hennie [45], and Kohavi [54].

In parallel, the area of switching circuit design and optimization was a focus of considerable attention. Boolean algebra provided the basic mathematical foundation for switching circuit design. Shannon [87] provided a basis for the synthesis of two-terminal switching circuits in 1949. Quine in 1952 [77] proposed the notion of prime implicants which is fundamental to the minimization of sum-of-product Boolean expressions Based on the work of Quine, McCluskey in 1956 [68] developed a systematic covering procedure to find the sum-of-products expression with a minimum number of product terms for any given Boolean function − the well-known Quine-McCluskey procedure for two-level logic minimization. Multilevel switching circuits, that contain more logic levels than the two levels in sum-of-product expressions, were also being designed and optimized. Work by Ashenhurst [6], Karp [79] and Lawler [58] in the late '50s and early '60s dealt with the minimization of multilevel Boolean graphs and circuits.

The current era of two-level and multilevel combinational logic synthesis began in the '70s with the advent of sophisticated two-level logic minimizers like MINI, developed at IBM, and multilevel logic synthesis systems developed by Muroga [19] and Davidson [23] among others. The first industrial logic synthesis system appears to be the LSS system developed at IBM, in the early '80s. Today, there are several university and commercial logic synthesis systems that have been used to design VLSI circuits. The efforts of Brayton [15] and other researchers have resulted in a large body of theoretical work in switching theory as well as efficient and practical combinational logic optimization algorithms.

As switching circuit design matured into combinational logic synthesis, the problems of finding minimal realizations of finite automata, which are implemented as switching circuits with feedback, subtly changed, and became more complex. The FSM optimization algorithms developed between the '50s to the '70s exclusively operated at the symbolic State Transition Graph (STG) level. These algorithms had to now predict the effects of a succeeding combinational logic minimization step, that was growing increasingly sophisticated.

Let us define an optimal encoding for a FSM as being one that results in *minimized* combinational circuitry that requires the smallest area. If the target implementation is a two-level circuit, and the combinational logic minimization algorithm used is restricted to distance-1 merging of product terms in the encoded FSM, finding an optimal encoding is relatively easy, and optimum algorithms run in reasonable times, even for large circuits! However, what we *now* have to model during the encoding step corresponds to full-fledged two-level Boolean minimization, or multilevel logic optimization, depending on the target implementation.

Exact two-level Boolean minimization algorithms have been known since the '50s. The methodology of using discrete "off-the-shelf" logic-gates was common till the early '70s, and therefore the problem of modeling two-level minimization during symbolic encoding was largely ignored, until the advent of VLSI. A primary focus of this book has been to document efforts in the predictive modeling of two-level and multilevel combinational logic minimization during symbolic encoding — a step that is key to the efficient synthesis of FSMs. Chapters 3 through 6 of this book deal with predictive modeling in the context of optimal input, output, and state encoding and FSM decomposition. Hierarchical specifications of sequential circuits have become common in an effort to manage increasing design complexity. The generalization of optimization algorithms to operate on hierarchical specifications is non-trivial and has been the subject of recent research, motivating us to describe hierarchical optimization methods in Chapter 7.

Most of the optimization problems in switching and automata circuit design have been proven to be nondeterministic polynomial-time complete (NP-complete) or NP-hard. This implies that the best available algorithms that guarantee optimality will require CPU times that grow exponentially with the circuit size, in the worst case. With integrated circuits moving from Small Scale Integrated (SSI) circuits to Very Large Scale Integrated (VLSI) circuits in the last two decades, it is obvious that efficient, heuristic algorithms with reasonable average-case performance are sorely needed.

Heuristic algorithms form a basic part of any Computer-Aided Design (CAD) tool for the optimization of combinational or sequential circuits. However, the development of efficient heuristics tailored for a particular optimization problem, requires a deep understanding of the mathematical structure underlying the optimization problem. In this book, we have presented optimum algorithms for FSM optimization problems, because they shed light on the mathematical structures underneath the problems. Moreover, successful heuristic procedures have been developed for most of these problems, based on the optimum algorithms. We have not presented every nuance, variation or implementation detail in the heuristic procedures developed for a particular problem, but instead have focused on the essential aspects of successful heuristic algorithms for each of the problems considered. We have also leaned toward the algorithmic, rather than the empirical analysis of the relative merits and demerits of the procedures described. Some of the heuristic techniques presented may become out-of-date in a few years and others may stay, but we believe that the theoretical results presented here will be long-lasting.

Acknowledgements

Over the years, several people have helped to deepen our understanding of automata theory, sequential logic synthesis, or VLSI CAD in general. We thank Jonathan Allen, Robert Brayton, Gaetano Borriello, Raul Camposano, Tim Cheng, Aart De Geus, Giovanni De Micheli, Alfred Dunlop, Abhijit Ghosh, Kurt Keutzer, Michael Lightner, Bill Lin, Tony Ma, Sharad Malik, Rick McGeer, Paul Penfield, Richard Rudell, Alexander Saldanha, Alberto Sangiovanni-Vincentelli, Fabio Somenzi, Albert Wang, Jacob White, and Wayne Wolf.

Lastly, we thank our families for their continual encouragement and support.

SEQUENTIAL LOGIC SYNTHESIS

Chapter 1

Introduction

1.1 Computer-Aided VLSI Design

Computer-Aided Design (CAD) of microelectronic circuits is concerned with the development of computer programs for the automated design and manufacture of integrated electronic systems, with emphasis today on the design and implementation of Very Large Scale Integrated (VLSI) circuits. Automated VLSI design is referred to as *VLSI synthesis.* Synthesis of VLSI circuits involves transforming a specification of circuit behavior into a mask-level layout which can be fabricated using VLSI manufacturing processes, usually via a number of levels of representation between abstract behavior and mask-level layout. Optimization strategies, both manual and automatic, are vital in VLSI synthesis in order to meet required specifications. However, the optimization problems encountered in VLSI synthesis are typically nondeterministic polynomial-time (NP)-complete or NP-hard [35]. Therefore, solutions to the optimization problems incorporate heuristic strategies, the development of which requires a thorough understanding of the problem at hand. Thus, optimization-based VLSI synthesis has evolved into a rich and exciting area of research.

Direct application of synthesis in industry has been a significant driving force for research in CAD. Simple marketing principles dictate that, other factors being equal, a product available sooner would capture a larger share of the market and would remain in use longer. The desire to reduce the time to design and manufacture has led to

the initial investment of considerable money and effort into the development of CAD tools capable of producing designs competitive with the
best manual designs. Today, constantly shrinking geometries and increasingly reliable manufacturing processes have led to complex systems
being implemented on a single chip, making the use of CAD tools com-
monplace and mandatory. In its turn, the rapid automation of the VLSI
design phase has allowed companies to keep pace with advances in other
areas of VLSI like computer architecture and manufacturing, leading to
a symbiotic relationship. As a consequence of this rapid development in
VLSI technology, it is currently possible to produce application-specific
integrated circuits (ASICs), microprocessors and other types of circuits
that contain millions of transistors.

1.2   The Synthesis Pipeline

There are several steps in the synthesis of mask-level layout
descriptions from specifications of circuit behavior. *Behavioral synthesis*
begins with a programming-language-like description of the functionality
and converts it to a register-transfer-level (RT-level) description that
implements the desired functionality. Among the issues involved at this
stage are the temporal scheduling of operations and the allocation of
hardware. For instance, decisions regarding the number of arithmetic
units in a digital signal processor are made in this step.

The steps involved in the transformation from the RT-level to
a gate-level circuit are collectively known as *logic synthesis*. Switching
and automata theory form the cornerstones of combinational logic synthesis and sequential logic synthesis respectively (or finite state machine
synthesis). Even though obtaining *some* gate-level circuit from a RT-
level description is straightforward, it is nontrivial to obtain a gate-level
circuit that conforms to the desired specifications. The first-order optimization criteria in this process are typically all or a desired subset of
area optimality, speed and testability.

Once the gate-level circuit has been obtained, mask-level layout
is derived using *layout synthesis* (physical design) tools. The physical
design styles of choice are typically gate array, sea-of-gates, standard-
cell and programmable gate array. Programmable gate array is popular
for extremely rapid prototyping of designs. Standard-cell is the design

style of choice in mostly-custom designs like microprocessors where only a portion of the chip is synthesized automatically. Gate array and sea-of-gates offer superior performance and integration density compared to programmable gate array and standard-cell design styles and are chosen when the complete chip is synthesized automatically.

1.3 Sequential Logic Synthesis

Almost all VLSI circuits are *sequential circuits i.e.* they contain memory or storage elements in the form of flip-flops or latches as well as combinational (or switching) circuitry. An RT-level description can be implemented using either synchronous or asynchronous sequential logic. While asynchronous design has certain advantages, design automation for the reliable asynchronous implementation of complex functionality is still in its infancy. The synchronous design paradigm is followed throughout this book.

Considerable progress has been made in the understanding of combinational logic optimization in the recent past and consequently a large number of university and industrial CAD programs are now available for the optimal synthesis of combinational circuits [8, 10, 22, 40]. These optimization programs produce results competitive with manually-designed logic circuits. For a review of current combinational optimization techniques, the reader is referred to [13, 14].

The understanding of sequential circuit optimization, on the other hand, is considerably less mature. The presence of internal state adds considerably to the complexity of the optimization problem. While the primary inputs and outputs are typically binary vectors at the RT-level, internal states are represented in symbolic form.

Sequential circuits are most often modeled using Finite State Machines (FSMs). A FSM is a mathematical model of a system (in our case, a switching circuit) with discrete inputs, discrete outputs and and a finite number of internal configurations or states. The state of a system completely summarizes the information concerning the past inputs to the system that is needed to determine its behavior on subsequent inputs. It is convenient to visualize a FSM as a directed graph with nodes representing the states and the edges representing the transitions between states. Such a graph is known as a State Transition

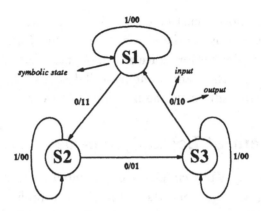

Figure 1.1: An example of a State Transition Graph (STG)

Graph (STG). An edge in the STG is labeled by the input causing the transition and the output asserted on the transition. A FSM can also be equivalently represented in tabular form by a State Transition Table (STT), each row of which corresponds to an edge in the STG. To deal with complexity, VLSI circuits are invariably specified in a hierarchical fashion. Large sequential circuits are typically modeled by smaller, interacting FSMs.

Synthesis tools are required to encode the internal symbolic states of FSMs as binary strings. This encoding determines the complexity and the structure of the sequential circuit which realizes the FSM, and therefore has a profound effect on its area, testability and performance. Stated differently, synthesis tools have the freedom of encoding states in such a way that the design constraints are satisfied. The notion of structure is generally associated with the manner in which a machine can be realized from an interconnection of smaller component machines as well as with the functional interdependences of its state and output variables. It may be desirable, for example, to construct the circuit with the minimum amount of logic, or to build it from an interconnection of smaller circuits to obtain superior performance.

A second degree of freedom available to sequential synthesis tools is based on the fact that the STG corresponding to a given functionality is not unique; transformations to the STG like state splitting, state merging or STG partitioning enable moving from one STG to another without changing the functionality. These transformations guide the encoding of states in a particular direction, in many cases into directions that would not have been taken otherwise. Such transformations are sometimes necessary to achieve the desired objectives.

1.4   Early Work in Sequential Logic Synthesis

Work in sequential logic synthesis dates back to the late '40s and '50s when discrete off-the-shelf components (relays and vacuum tubes) were used. In the 1960's, small-scale integrated circuits(SSI) became popular and much of the work in that period was motivated by the need to reduce the number of latches in the circuit since that meant a reduction in the number of relatively expensive chips on the circuit board. Also, since combinational logic synthesis was still in its infancy, techniques for state encoding and FSM decomposition were unable to target the combinational logic complexity of the sequential circuit effectively.

The minimization of the number of states in completely specified FSMs was first investigated by Moore [75], Huffman [48, 49] and Mealy [70]. This work was later extended to the reduction of states in incompletely specified machines by Ginsburg [39] and Unger [76]. An interesting technique for deriving maximal compatibles in state reduction using Boolean algebra was reported in a short communication by Marcus [67].

The relationship between state encoding and the structure of the resulting sequential circuit was first investigated in terms of the algebraic theory of partitions by Hartmanis [41, 42] and later by Hartmanis and Stearns [44, 89]. Contributions to machine-structure theory were also made by Karp [50], Kohavi [53], Krohn and Rhodes [55], Yoeli [98, 99] and Zeiger [100]. The concept of state splitting to augment the possibilities of finding desirable decompositions and state assignments was developed, among others, by Hartmanis and Stearns [44], Zeiger [100] and Yoeli [98]. The book by Hennie [45] provides a lucid and intuitive description of the above contributions.

State encoding was treated from a different point of view by Armstrong [1] and Dolotta and McCluskey [32]. The procedure of Armstrong formulated the state encoding problem as one of assigning codes so that pre-derived adjacency relationships between states are satisfied in the Boolean domain. To a certain extent, the procedure of [1] inspired some state assignment algorithms (*e.g.* Devadas *et al* [26]) developed recently for targeting multilevel implementations.

1.5   Recent Developments

1.5.1   State Encoding

Considerable work has been done in the area of sequential logic synthesis in the last five years (*e.g.* [26, 73, 74, 85]). An important development in state encoding was the step from predictive to exact approaches for state encoding targeting two-level implementations. A fundamental result which made this possible was the establishment of a link between the size of a minimized symbolic tabular representation of a FSM and the maximum number of product terms required in a Programmable Logic Array (PLA) implementing the same FSM after encoding the states in the work by De Micheli *et al* [73]. The approach followed in [73] involves a two-step process. In the first step, the STT of the FSM is symbolically minimized using a two-level multiple-valued minimization program like ESPRESSO-MV [81]. This minimization step generates constraints that the state codes must satisfy if the PLA resulting from the encoding is to have as small or an equal number of product terms as the minimized STT. Obtaining a state encoding that satisfies the constraints forms the second step of the procedure. Since multiple-valued input by itself cannot account for the interactions in the next-state plane of the encoded FSM, the approach of [73] effectively approximated the state encoding problem as one of input encoding.

States appear both in the input and output planes of the PLA and therefore state encoding is actually an input-output encoding problem. The interactions between product terms in the next-state plane can occur either due to dominance or disjunctive relationships between state codes. If the code for state s_1 dominates the code for state s_2 (*i.e.* the code for s_1 has 1's in all positions that the code for s_2 has 1's), the input parts of the cubes asserting the next-state s_1 can be used as don't cares in order to optimize the cubes asserting the next-state s_2. Similarly, if the code for state s_1 is the disjunction of the codes for states s_2 and s_3, the input part of the cubes that assert the next-state s_1 can be optimized using the input parts of the cubes asserting the next-states s_2 and s_3.

Work by De Micheli [71] was the first to take advantage of interactions between product terms in the next-state plane. In particular, it attempted to maximize the cardinality reductions due to dominance relationships between state codes by reducing the problem to one of heuristically finding the order in which states should be encoded.

Further understanding of interactions in the output plane led to the work of Devadas and Newton [30] in which a procedure was presented for encoding symbolic outputs to achieve the minimum product-term count. By means of this procedure, relationships between codes due to both dominance and disjunctive relationships in the output plane can be handled simultaneously. The search for all possible relationships and their effects is carried out by modifying the classical Quine-McCluskey method [68, 78] for prime implicant generation and covering. The notion of Generalized Prime Implicants (GPIs) was introduced in [30] for that purpose. GPIs correspond to a weaker form of primality than conventional prime implicants in that for the same symbolic cover, the set of GPIs contains the set of conventional prime implicants and is much larger than it. The advantage of GPIs is that they allow interactions in the output plane to be handled formally. Devadas and Newton [30] also extended the GPI formulation to arrive at an exact method for state encoding, which guarantees minimum product-term count in a resulting two-level logic implementation. While the use of GPIs leads to an exact procedure, their use is only viable for small FSM examples.

One of the important steps in the encoding of symbolic states in a FSM is the satisfaction of encoding constraints. In the most general case, these constraints are comprised of both input and output constraints. In the case of input constraints, one is typically given a constraint matrix, A, each column of which corresponds to a state and each row to a constraint. The goal is to find an encoding matrix S where each row in S corresponds to the binary code chosen for a state, such that the constraints implied by A are satisfied and the number of columns in S is minimum. Satisfying the input constraints entails obtaining state codes such that for each row in A the states present in the row form a face in the Boolean n-space and the states absent from the row are excluded from that face. This problem is called the *face-embedding* problem. Output constraints, on the other hand, force bitwise dominance and disjunctive relationships between the state codes.

Finding the minimum-length encoding satisfying the constraints is an NP-hard problem [83]. De Micheli *et al* [73] provided a number of results for reducing A without violating the original set of constraints and proposed a row-based encoding algorithm. According to this algorithm, S is constructed row by row with the invariant that the constraints corresponding to the rows of A already seen are not violated by the portion of S constructed thus far. Columns are added to S when necessary. This algorithm was found to be effective only for small examples. Column-based algorithms were proposed by De Micheli [71] and Devadas *et al* [31]. In a column-based algorithm, S is constructed one column (one bit) at a time. None of these algorithms guarantee a minimum-length encoding. The work by Villa *et al* [94] includes an algorithm for obtaining the minimum-length encoding satisfying all input constraints. This work represents a refinement of the methods developed previously [71, 73].

An alternate approach to constraint satisfaction uses the notion of dichotomies. A dichotomy is defined as a disjoint two-block partition on a set, in our case the set of states. The notion of dichotomies was first introduced for hazard-free state encoding of asynchronous circuits by Tracey [92]. Dichotomies were revisited for constrained state encoding by Yang and Ciesielski [97]. The main result of that work was to show that the minimum number of bits required to encode a set of constraints is equal to the minimum number of prime dichotomies required to cover all the seed dichotomies. An implementation based on graph coloring was suggested for this approach. An alternate implementation of the dichotomies approach based on prime generation and classical unate covering [80] is due to Saldanha *et al* [83]. It was also shown in that work how output constraints could be handled using dichotomies. In this case, the generation of primes was carried out using the same approach as used by Marcus [67] for generating maximal compatibles in state minimization.

In other work on constraint satisfaction [30], it was shown that if the encoding length is known apriori, all constraints can be represented by Boolean equations. While this idea is attractive in theory, a naive representation of all constraints involved as Boolean equations can lead to an intractable satisfiability problem. Simulated annealing has also been successfully used for constraint satisfaction [26, 61, 94].

State encoding targeting multilevel implementations is an even more difficult problem. The main reason for this is that combinational optimization of multilevel circuits is itself not an exact science. Even so, certain optimization strategies like common factor extraction are fundamental to multilevel optimization and a number of predictive state encoding procedures have been proposed that maximize the gains due to these basic strategies. Devadas *et al* [26] proposed an algorithm for state encoding in which the likelihood of finding common subexpressions and common cubes in the logic prior to optimization was enhanced by minimizing the distance in the Boolean space between chosen states. A similar approach was followed in [61, 95]. Recently, attempts have been made to extend the encoding paradigm followed for two-level circuits to multilevel circuits. In the work of Malik *et al* [56] techniques were proposed for optimizing multilevel circuits with multiple-valued input variables. As in the two-level case, these optimizations lead to constraints on possible binary encodings for the multiple-valued variables that must be satisfied if the effects of the optimizations are to be preserved after encoding.

1.5.2   Finite State Machine Decomposition

FSM decomposition can be used to obtain partitioned sequential circuits with the desired interconnection topology. Sequential circuit partitioning can lead to improved performance, testability and ease of floor-planning. Decomposition at the STG-level allows a larger solution space to be searched for partitioning sequential circuits than techniques that operate at the logic level. However, the drawbacks of the above approaches which operate at the STG level is that it is difficult to accurately predict the effect of an operation at the symbolic level on the cost of the resulting logic.

In recent work on FSM decomposition, it was recognized by Devadas and Newton [29] that many FSMs possess isomorphic subgraphs in their STGs. The implementation of multiple instances of such isomorphic subgraphs (called *factors* in [29]) as a single separate submachine distinct from the parent machine can lead to reduced area and improved performance. The interactions between the decomposed submachine and

the factor submachine are typically bi-directional and therefore corresponds to a general decomposition, as opposed to a cascade decomposition where the interaction is uni-directional (a head submachine drives
a tail submachine). The authors also demonstrated that this decomposition approach leads to an encoding strategy that takes advantage
of some interactions in the next-state plane. Work in the cascade decomposition of sequential machines required the detection of non-trivial
preserved covers or partitions over the set of states in the machine. Similarly, in order find a good general decomposition of a machine, one has
to find large isomorphic subgraphs in the given State Transition Graph.

The notion of generalized prime implicants introduced in [30]
has been extended to FSM decomposition [4]. This allows for the development of an exact algorithm for arbitrary topology, multi-way, FSM
decomposition. The algorithm targets a cost function that corresponds
to a weighted sum of the number of product terms in the encoded and
two-level minimized submachines. This cost function is more accurate
at predicting logic-level complexity than the cost functions targeted by
earlier approaches to decomposition.

1.5.3 Sequential Don't Cares

The methods described so far begin with a single or lumped
symbolic State Transition Graph (STG). The initial specification of a
sequential circuit, as specified by a designer may in fact be an interconnection of several machines, each of which is described by a separate
STG. While it is possible to "flatten" the given hierarchical description
into a lumped STG, doing so may require exorbitant amounts of CPU
time or memory, since the lumped STG may have too large a number
of states. The use of state encoding techniques to *independently* encode
each of the constituent machines in the design may result in suboptimal solutions. A primary reason for this suboptimality is that don't
care conditions may exist at the boundaries of any two interacting machines, and these don't cares have to be exploited in order to obtain an
optimized, high-quality design.

Single-vector don't care inputs to a sequential machine result
in a STG that is incompletely specified. The utilization of these don't
cares in state minimization has received considerable attention. Early
work in sequential don't cares [51, 93] also recognized the existence of

input don't care sequences (rather than single vectors) to a tail machine when it is being driven by a head machine. An example is given in [93] that shows that the number of states in the tail machine can be reduced by utilizing these input don't care sequences. The procedure described by Kim and Newborn [51] ensures state minimality under all possible don't care input sequences using methods similar to those used in cascade decomposition [44, 45].

Recently, more efficient methods to exploit sequential don't cares were developed [24], and a new class of don't cares, namely output don't care sequences were shown to exist. Output don't care sequences correspond to two or more output sequences from a head machine that elicit the same response from a tail machine. Optimization of sequential circuits under these don't cares is currently a subject of extensive research.

1.5.4 Sequential Resynthesis at the Logic Level

The observation that significant gains could be accrued by optimizing certain sequential circuits at the logic-level was made by Leiserson, Rose and Saxe [59] for systolic arrays. Systolic arrays are sequential circuits which are capable of operating at very high clock frequencies because they are designed as highly pipelined structures with very little logic between pipeline latches on any path in the network. This design methodology implies that the number of latches is usually very large. Given an initial design, the problem of retiming or relocating these latches so that the number of latches is minimized, with the circuit still satisfying the clock frequency requirement and with the functionality of the circuit remaining unchanged, was formulated as an integer-linear programming problem, that is solvable in polynomial time, in [59]. The work of Malik *et al* [66] extended the ability to retime latches to a larger class of circuits. In that work, the general problem of state encoding was reduced to one of latch retiming. While this notion is certainly attractive in theory, it was found that most sequential circuits implementing control-type functions are not amenable to global optimization by latch retiming because of the tight feedback paths they possess. The reason that latch retiming was successful for systolic arrays was that most of the latches are used as pipeline latches and very few of them are in unique feedback loops.

Techniques to efficiently extract State Transition Graph (STG) descriptions from logic-level implementations have been proposed recently [5]. Note that the extracted STG already has an encoding of states. However, this encoding may or may not be an efficient encoding. The encoding can be "thrown away" and a symbolic STG can be obtained from the extracted STG. Given this symbolic STG, the various encoding and decomposition procedures described can be applied. The efficiency of this procedure is related to the fact that detection of some equivalent states and the detection of edges that can be combined in a two-level representation of the STG is performed during the extraction process itself. As a result, this procedure avoids having to extract an intermediate representation in which the fanout of equivalent states are explicitly enumerated and edges that could have been been combined are enumerated separately. This procedure is akin to test generation algorithms — in fact, variations of this procedure have been used in sequential logic testing and verification. The procedure and its variations have been described in a companion book [38].

Exploiting don't care conditions at the logic level is also a subject of current research. The effect of exploiting don't care conditions at the logic level on the structure of the STG has been explored recently [3, 27]. Insight into the phenomenon of state splitting (having equivalent states) in the STG producing better logic implementations than a reduced STG can be obtained by understanding the meaning of the different don't care conditions. Further, since these don't care optimization methods work directly on the logic-level representation, they alleviate the problems of predictive state minimization and encoding. There are also intimate relationships between sequential don't cares and stuck-at fault redundancy in interacting FSMs.

1.6 Organization of the Book

The basic terminology used in the book is defined in Chapter 2.

Chapter 3 is devoted to input encoding. The use of multiple-valued minimization provides an exact solution to the input encoding problem. Various input encoding methods based on the generation of encoding constraints via multiple-valued minimization are described in Chapter 3. Strategies to satisfy the resulting input constraints using a minimal number of encoding bits are also described. Finally, recent work in input encoding targeting multilevel logic implementations is described at the end of the chapter.

Output encoding is the subject of Chapter 4. Early heuristic approaches to output encoding are first described. The notion of generalized prime implicants (GPIs) provides a formalism for an exact method to solve the output encoding problem. The various steps in this procedure are described in detail in Chapter 4.

State assignment is an input-output encoding problem, since the symbolic State Transition Graph has both a symbolic input (the present state field) as well as a symbolic output (the next state field). The integration of the approaches presented in Chapters 3 and 4 to solve the state assignment problem is the subject of Chapter 5. In particular, we describe some efficient heuristic methods for state assignment, as well as the integration of the exact output encoding method and multiple-valued input minimization to solve the state assignment problem exactly.

Chapter 6 is devoted to finite state machine decomposition. Early work in FSM cascade decomposition is first reviewed in Chapter 6. An approach to general decomposition based on factorization is then described. An approach to FSM decomposition based on generalized prime implicants is presented in Chapter 6 which makes it possible to target logic-level optimality of the partitioned sequential circuit. For the cost function chosen, the algorithm presented in this chapter can be used to obtain a decomposition with minimum cost. It is shown in this chapter that state encoding and FSM decomposition are two sides of the same coin. FSM decomposition can be viewed as a structural transformation of the FSM that guides the subsequent steps of state encoding in the desired direction. Based on this premise, one state encoding strategy involves decomposing the FSM prior to performing the actual encoding on the resulting smaller submachines. Both exact and heuristic methods for FSM decomposition that target logic-level complexity are presented in Chapter 6.

Sequential don't care conditions, corresponding to sequences that never occur at the inputs to a machine, or sequences that produce the same response from a given machine are the subject of Chapter 7. Two basic types of sequential don't cares are introduced, and methods for exploiting these don't cares to minimize the number of states in the constituent machines of a hierarchical specification are described. The notion of unconditional and conditional state compatibility is introduced in Chapter 7, which allows for the systematic classification of don't cares in arbitrary interconnections of FSMs, where the machines communicate via their present states. These don't cares can be used to optimize logic-level single or interacting FSMs.

Chapter 8 concludes the book. Optimization only represents one facet of VLSI design. Sequential testing and verification represent two other facets in the design of VLSI sequential circuits. Exploring the inter-relationships between these facets is evolving into a rich and rewarding area of CAD research. In Chapter 8 directions for future research are indicated.

Most of the ideas presented in this book have been implemented in the sequential logic synthesis system FLAMES. Implementation details of FLAMES can be found in [2].

Chapter 2

Basic Definitions and Concepts

Most of the terminology used in this book is standard and in common use in the synthesis community [54, 13, 14]. This chapter is devoted to the definition of the nontrivial terminology and an elucidation of some of the basic concepts.

2.1 Two-Valued Logic

A **binary variable** is a symbol representing a single coordinate of the Boolean space (*e.g.* a). A **literal** is a variable or its negation (*e.g.* a or $\bar{a}$). A **cube** is a set C of literals such that $x \in C$ implies $\bar{x} \notin C$ (*e.g.*, $\{a, b, \bar{c}\}$ is a cube, and $\{a, \bar{a}\}$ is not a cube). A cube (sometimes called a **product term**) represents the conjunction, *i.e.* the Boolean product of its literals. The trivial cubes, written 0 and 1, represent the Boolean functions 0 and 1 respectively. A **sum-of-products expression** is the disjunction, *i.e.* a Boolean sum, f, of cubes. For example, $\{\{a\}, \{b, \bar{c}\}\}$ is an expression consisting of the two cubes $\{a\}$ and $\{b, \bar{c}\}$.

A cube may also be written as a bit vector on a set of variables with each bit position representing a distinct variable. The values taken by each bit can be 1, 0 or 2 (don't-care), signifying the true form, negated form and non-existence respectively of the variable corresponding to that position. A **minterm** is a cube with only 0 and 1 entries. Cubes can be classified based on the number of 2 entries in the cube. A cube with

1–00	(inp1, inp3)	out3	100		1–00	101	001	100
111–	(inp1, inp3)	out3	011		111–	101	001	011
101–	(inp1, inp2)	out2	110		101–	110	010	110
1–01	(inp1, inp3)	out1	101		1–01	101	100	101
0–––	(inp1)	out1	111		0–––	100	100	111

(a)   (b)

Figure 2.1: An example of a symbolic cover and its multiple-valued representation

k entries or bits which take the value 2 is called a k-**cube**. A minterm thus is a 0-**cube**. A cube c_1 is said to **cover (contain)** another cube c_2, if c_1 evaluates to 1 for every minterm for which c_2 evaluates to 1. A **super-cube** of a set of cubes is defined as the smallest cube containing all the minterms contained in the set of cubes.

The **on-set** of a function f is the set of minterms for which the function evaluates to 1, the **off-set** of f is the set of minterms for which f evaluates to 0, and the **don't-care set** or the **DC-set** is the set of minterms for which the value of the function is unspecified. A function which does not have a DC-set is a **completely-specified** function. A function with a non-empty DC-set is termed **incompletely-specified**.

An **implicant** of f is a cube that does not contain any minterm in the off-set of f. A **prime implicant** of f is an implicant which is not contained by any other implicant of f, and which is not entirely contained in the DC-set.

A **two-level cover** for a Boolean function is a set of implicants which cover all the minterms in the on-set, and none of the minterms in the off-set. A cover is **prime** if it composed entirely of prime implicants. A cover is **irredundant** if removing any single implicant results in a set of implicants that is not a cover. A **minimum cover** is a cover of minimum cardinality over all the possible covers for the function.

A minterm m_1 is said to **dominate** another minterm m_2 if for each bit position in which m_2 has a 1, m_1 also has a 1. If m_1 dominates m_2, we write $m_1 \succ m_2$ or $m_2 \prec m_1$.

The **distance** between two minterms is defined as the number of bit positions they differ in.

2.2   Multiple-Valued Logic

In general, a logic function may have **symbolic** (also known as **multiple-valued**) input or output variables in addition to binary variables. Like binary variables, a symbolic variable also represents a single coordinate, with the difference that a symbolic variable can take a subset of values from a set, say P_i, that has a cardinality greater than two.

Let the symbolic variable v take values from $V = \{v_0, v_1, \ldots, v_{n-1}\}$. v may be represented by a multiple-valued (MV) variable, X, restricted to $P = \{0, 1, \ldots, n-1\}$, where each symbolic value of v maps onto a unique integer in P.

Let $B = \{0, 1\}$. A binary valued function $\mathbf{f}$, of a single MV-variable X and $m-1$ binary-valued variables, is a mapping: $\mathbf{f} : P \times B^{m-1} \rightarrow B$. Each element in the domain of the function is called a **minterm** of the function. Let $S \subseteq P$. Then X^S represents the Boolean function:

$$X^S = \begin{cases} 1 & \text{if } X \in S \\ 0 & \text{otherwise} \end{cases}$$

X^S is called a **literal** of variable X. If $|S| = 1$ then this literal is also a minterm of X. For example, $X^{\{0\}}$ and $X^{\{0,1\}}$ are literals and $X^{\{0\}}$ a minterm or a 0-cube of X. A minterm or 0-cube over a set of inputs is a cube in which all variables take only a single value. If $S = \phi$, then the value of the literal is always 0. If $S = P$ then the value of the literal is always 1. For these two cases, the value of the literal may be used to denote the literal. We note the following:

1. $X^{S_1} \subseteq X^{S_2}$ if and only if $S_1 \subseteq S_2$.

2. $X^{S_1} \cup X^{S_2} = X^{S_1 \cup S_2}$.

3. $X^{S_1} \cap X^{S_2} = X^{S_1 \cap S_2}$.

As in the two-level case, the literal of a binary-valued variable y is defined as either the variable or its Boolean complement. A **cube** or a **product term** is a Boolean product (AND) of literals. If a product term evaluates to 1 for a given minterm, it is said to contain the minterm.

When an output variable, say Y_j, is symbolic it implies that Y_j can take a value from a set P_j, of values when the input is some minterm. An example of a symbolic cover with one symbolic input and one symbolic output is shown in Figure 2.1(a). The symbolic input or output variable takes on **symbolic values**. In the example, the symbolic input variable takes values from the set $\{inp1, inp2, inp3\}$ while the symbolic output variable takes a value from the set $\{out1, out2, out3\}$. A convenient method for representing a symbolic variable that can take values from a set of cardinality n is to use an n-bit vector to depict a literal of that variable such that each position in the vector corresponds to a specific element of the set. A 1 in a position in the vector signifies the presence of an element in the literal while a 0 signifies the absence. An example of a one-hot representation for the symbolic cover of Figure 2.1(a) is shown in Figure 2.1(b). For instance, the MV-literal 101 represents the union of the symbolic values $inp1$ and $inp3$, and $out1$ has been represented as 100. The binary-valued encoding of a symbolic value v_i is denoted $enc(v_i)$.

2.3   Multilevel Logic

A **Boolean network** η is a directed, acyclic graph [10] (DAG) such that for each node n_i in η there is an associated Boolean function f_i, and a Boolean variable y_i representing f_i. There is a directed edge from n_i to n_j if f_j explicitly depends on y_i or $\overline{y_i}$. Further, some of the variables in η may be classified as **primary inputs** or **primary outputs**. The Boolean network is a technology-independent multilevel representation of a completely-specified combinational logic circuit. The primary inputs represent the inputs to the circuit, and the primary outputs represent the outputs of the circuit.

2.4   Multiple-Valued Input, Multilevel Logic

A **sum-of-products** (SOP) is a Boolean sum (OR) of product terms. For example: $X^{\{0,1\}}y_1y_2$ is a product term, y_1y_2 a cube and $X^{\{0,1\}}y_1y_2 + X^{\{3\}}y_2y_3$ is a SOP. A function **f** may be represented by a SOP expression f. In addition **f** may be represented as a factored form. A factored form is defined recursively as follows.

Definition 2.4.1 *An* SOP *expression is a factored form. A sum of two factored forms is a factored form. A product of two factored forms is a factored form.*

$X^{\{0,1,3\}}y_2(X^{\{0,1\}}y_1 + X^{\{3\}}y_3)$ is a factored form for the SOP expression given above.

A logic circuit with a multiple-valued input is represented as a MV-network [56]. A MV-**network** η, is a directed acyclic graph (DAG) such that for each node n_i in η there is associated a binary-valued, MV-input function f_i, expressed in SOP form, and a Boolean variable y_i. There is an edge from n_i to n_j in η if f_j explicitly depends on y_i. Further, some of the variables in η may be classified as **primary inputs** or **primary outputs**. These are the inputs and outputs (respectively) of the MV-network. The MV-network is an extension of the Boolean network defined in the previous section to permit MV-input variables. Since each node in the network has a binary-valued output, the non-binary (MV) inputs to any node must be primary inputs to the network.

2.5 Finite Automata

A **finite state machine** (FSM), M, is a six-tuple $(Q, \Sigma, \Delta, \delta, \lambda, q_0)$, where Q is a finite set of states, Σ is a finite input alphabet, δ is the transition function mapping $Q \times \Sigma \rightarrow Q$, Δ is the output alphabet and q_0 is the initial or starting state.

We have two different types of finite state machines depending on the mapping function λ. In a **Moore** FSM λ is a mapping $Q \rightarrow \Delta$, whereas in a **Mealy** FSM λ is a mapping $Q \times \Sigma \rightarrow \Delta$. Thus, in a Moore FSM the output is associated with each state, and in a Mealy FSM the output is associated with each transition.

In a logic-level implementation of a FSM, a combinational logic network realizes the mappings δ and λ and feedback registers store state. Given a logic-level implementation of a FSM with N_i primary inputs and N_o primary outputs, the input and output alphabets are Boolean spaces of dimensions N_i and N_o respectively. If the FSM has N_b registers, then each state of the FSM is a minterm over a Boolean space of dimension N_b. The FSM can have a maximum of 2^{N_b} states.

A directed graph, called a **State Transition Graph**, $G(V, E)$, is associated with a FSM as follows. The vertices of the graph V correspond to the states of the FSM. If there is a transition from state q to state p on input a producing an output z, then there is an arc in E labeled a/z from vertex q to vertex p. The input symbol associated with this arc is usually represented by a minterm (specifying a combination of primary input values). Multiple arcs may exist between two states on different input minterms producing the same output. A subset (or all) of these arcs may be represented more compactly by a single arc whose input symbol is a cube which is the union of the corresponding minterms. For example, in a FSM with two primary inputs, two states s_1 and s_2 and a single output, we may have two transitions from state s_1 and s_2 on input symbols 10 and 11, each producing the output 1. In a more compact representation, we will have a *single* transition arc on the input symbol 1– producing the output 1.

The input combination and present-state corresponding to an edge, $e \in E$, or a set of edges is (i, s), where i and s are cubes. The **fanin** of a state, $q \in V$ is a set of edges and is denoted $fanin(q) \in E$. The **fanout** of a state $q \in V$ is also a set of edges and is denoted $fanout(q) \in E$. The output and the next state of an edge $(i, s) \in E$ are $o((i, s))$ and $n((i, s)) \in S$ respectively. The input, present state, next state and output of an edge $e \in E$ are also denoted $e.Input$, $e.Present$, $e.Next$ and $e.Output$ respectively.

Given N_i inputs to a machine, 2^{N_i} edges with minterm input labels fan out from each state. A STG where the next-state and output labels for every possible transition from every state are defined corresponds to a **completely-specified machine**. An **incompletely-specified machine** is one where at least one transition edge from some state is not specified.

A State Transition Table (STT) is a tabular representation of the FSM. Each row of the table corresponds to a single edge in the STG. Conventionally, the left most columns in the table correspond to the primary inputs and the right most columns to the primary outputs. The column following the primary inputs is the present-state column and the column following that is the next-state column.

A **starting** or **initial state** (also called the **reset state**) is assumed to exist for a machine. Given a logic-level description of a sequential machine with N_b flip-flops, 2^{N_b} possible states exist in the machine. A state which can be reached from the reset state via some input vector sequence is called a **valid state** (or a **reachable state**). The corresponding input vector sequence is called the **justification sequence** for that state. A state for which no justification sequence exists is called an **invalid state** (or an **unreachable state**).

A **differentiating sequence** for a pair of states q_1, $q_2 \in Q^2$ of a machine M is a sequence (or string) of input vectors such that the last vector produces different outputs when the sequence is applied to M, when M is initially in q_1 or when M is initially in q_2. Two states q_1, q_2 in a completely-specified machine M are **equivalent** (written as $q_1 \equiv q_2$), if they do not possess a differentiating sequence. If two states q_1, q_2 in an incompletely-specified machine M do not possess a differentiating sequence, they are said to be **compatible** (written as $q_1 \cong q_2$).

A **reduced** or **state minimal** machine is one where no pair of states are equivalent.

A State Transition Graph G_1 is said to be **isomorphic** to another State Transition Graph G_2 if and only if they are identical except for a renaming of states.

Chapter 3

Encoding of Symbolic Inputs

3.1 Introduction

Variables in the specification of a design are often multiple-valued or symbolic. However, in order to implement the design as a VLSI circuit, a binary encoding for the various symbolic values of the symbolic variable has to be chosen. Examples of symbolic variables include the "instruction" of a computer which takes values over the different instructions. In a controller implementation, each instruction is assigned a unique binary opcode. Another example is an arithmetic unit with a symbolic input in its specification that controls the operation the unit performs.

The focus of this chapter is the problem of determining an encoding for a function with symbolic inputs, such that the eventual implementation *after* encoding and logic minimization is of minimum size. Encoding problems are difficult because they have to model the complicated optimization step that follows. For instance, if we have symbolic truth-tables like those in Figure 2.1(a), which are to be implemented in Programmable Logic Array (PLA) form, one wishes to code the $\{inp1, inp2, inp3\}$ symbolic values so as to minimize the number of product terms (or the area) of the resulting PLA after two-level Boolean minimization. A straightforward, exhaustive search technique to find a optimum encoding with the minimum number of product terms would

```
0 s1   s2 1
1 s1   s4 0
0 s2   s2 1
1 s2   s1 1
0 s3   s3 0
1 s3   s4 0
0 s4   s2 1
1 s4   s1 1
```

Figure 3.1: A symbolic tabular representation of a Finite State Machine

require $O(N!)$ *exact* two-level Boolean minimizations, where N is the number of symbolic values. Two-level Boolean minimization algorithms are very well developed – the programs ESPRESSO-EXACT [80] and MC-BOOLE [21] minimize large functions exactly within reasonable amounts of CPU time. However, the *number* of required minimizations makes an exhaustive search approach to optimum encoding infeasible for anything but the smallest problems.

A major contribution in input encoding that modeled the subsequent step of two-level Boolean minimization was that of De Micheli, Brayton and Sangiovanni-Vincentelli [73] in which a new paradigm was proposed for input encoding. It was suggested that input encoding be viewed as a two-step process. In the first step, a tabular representation of the function with the symbolic input is optimized at the symbolic level. This optimization step generates constraints on the relationships between codes for different symbolic values. In the second step, states are encoded in such a way the constraints are satisfied. Satisfaction of the constraints guarantees that the optimizations at the symbolic-level will be preserved in the Boolean domain. If all the optimizations possible in the Boolean domain are also possible at the symbolic level, an optimal solution is guaranteed using this approach. This approach is described in detail in Section 3.2. Various methods for satisfying face embedding constraints are described in Section 3.3. Note that state assignment can be approximated as an input encoding problem by ignoring the next state field in a finite state machine description like the one shown in Figure 3.1.

De Micheli, Brayton and Sangiovanni-Vincentelli [73] used the above paradigm successfully to solve the input encoding problem when the target implementation was two-level logic. However, control logic is more often than not implemented as multilevel logic, and the optimization step following encoding is one of multilevel logic optimization. While two-level Boolean minimization is an exact science, multilevel minimization is not yet well characterized. Thus, the problem of input encoding targeting multilevel logic implementation proved much more difficult than its two-level counterpart. An initial effort in this direction was that of Devadas *et al* [26], where an attempt was made to model the common cube extraction [14] step in multilevel logic synthesis at the symbolic level, for the input encoding and state encoding problems. Another effort in input and state encoding targeting multilevel logic implementations focused on modeling the more powerful common kernel extraction step in multilevel minimization at the symbolic level [95]. We will defer the description of these approaches to Chapter 5 where approaches to state assignment are described.

Recently, substantial progress has been made in applying the paradigm of symbolic minimization following by constrained encoding to the input encoding problem targeting multilevel logic. In particular, Malik *et al* [56, 65] have defined notions of multiple-valued, multilevel minimization, which allow accurate modeling of operations in multilevel optimization for the input encoding problem. We describe this approach in Section 3.4.

3.2   Input Encoding Targeting Two-Level Logic

The initial effort in input and state encoding using the two-step paradigm targeted optimal two-level implementations of encoded FSMs. In this section, we will be concerned with finding an encoding for the symbolic values of a symbolic input variable such that the resulting two-level implementation has a minimum number of product terms.

Consider the function with a symbolic input of Figure 3.2(a). Two different encodings for the function are shown in Figure 3.3(a) and 3.3(b), respectively. One results in five product terms, the other results in four. We are interested in an algorithmic means of predicting the cardinality of the encoded and minimized function. To this end, we describe one-hot coding and multiple-valued minimization in Section 3.2.1

```
10 inp1   01              10 inp1   01
01 inp1   10              01 inp1   10
1- inp2   01              1- inp2   01
01 inp2   10              01 inp2   10
11 inp3   01              11 inp3   01
01 inp3   10              01 inp3   10
-- inp4   11
    (a)                       (b)
```

Figure 3.2: A function with a symbolic input

```
inp1 → 10    10 1-  01
inp2 → 01    01 --  10        inp1 → 00    11 -1  01
inp3 → 00    11 0-  01        inp2 → 01    10 0-  01
inp4 → 11    1- -1  01        inp3 → 11    01 --  10
             -- 11  11        inp4 → 10    -- 10  11
      (a)                            (b)
```

Figure 3.3: Two different encodings for the symbolic input

and face embedding constraints that can be used to guide an optimal encoding process in Section 3.2.2.

3.2.1   One-Hot Coding and Multiple-Valued Minimization

A one-hot coding for the symbolic variable of the function of Figure 3.2(b) is shown in Figure 3.4(a). The symbolic value *inp1* has been coded with 100, *inp2* with 010 and *inp3* with 001. The unused codes have been added as don't care conditions. Minimizing the one-hot coded function with the associated don't care set results in the three product term result of Figure 3.4(b).

The first observation made in [73] was that given an arbitrary function with multiple symbolic inputs, one-hot coding each of the symbolic inputs, and exactly minimizing the encoded function with the associated unused code don't cares, was guaranteed to result in a minimum product term implementation. This is because any two implicants in the symbolic cover that can potentially merge under some encoding of symbolic values, will also be mergeable under a one-hot code. For instance, in our example of Figure 3.4, *inp1* coded with 100 and *inp2* coded with

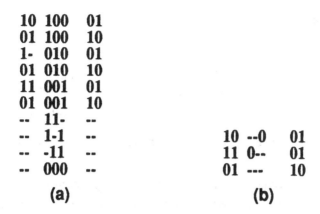

```
10 100   01
01 100   10
1-  010   01
01 010   10
11 001   01
01 001   10
--  11-   --
--  1-1   --                 10 --0   01
--  -11   --                 11 0--   01
--  000   --                 01 ---   10
    (a)                          (b)
```

Figure 3.4: One-hot coding and minimization

010 have merged into the first product term of the minimized cover, namely $- - 0$ due to the specified don't care set.

Given the above observation, the input encoding problem can be solved exactly when the objective is minimum product term count, simply by one-hot coding the given function, and applying two-level Boolean minimization with an appropriate don't care set. Of course, a one-hot code may require too many bits to be practical. The major contribution in [73] was to provide a theory for the general encoding of symbolic inputs while guaranteeing minimum cardinality of the encoded implementation.

A problem with the application of a two-level Boolean minimizer for large input encoding problems is that the don't care sets involved can be very large. In fact, for symbolic inputs with over 20 symbolic values, binary-valued minimizers like ESPRESSO [13] require too long a time to minimize the one-hot coded function. An alternate strategy is to use a multiple-valued minimizer like ESPRESSO-MV [81] in order to avoid explicit minimization under a large don't care set.

In Figure 3.5(a), we have replaced each symbolic value by a minterm of a multiple-valued variable. The multiple-valued variable has been represented in one-hot form. While Figures 3.4 and 3.5 look identical, the meaning of the 2nd column is quite different. In Figure 3.4, we have five binary-valued input variables. We have a don't care set associated with combinations of the last three variables not occurring. In Figure 3.5, we have two binary-valued input variables, and one three-

```
10 100  01
01 100  10
1- 010  01
01 010  10      10 110  01      10 (inp1, inp2)            01
11 001  01      11 011  01      11 (inp2, inp3)            01
01 001  10      01 111  10      01 (inp1, inp2, inp3)      10
     (a)             (b)                    (c)
```

Figure 3.5: Multiple-valued minimization

valued input variable. Given the multiple-valued algebra of Section 2.2, any combination of minterms of the multiple-valued variable can merge into a multiple-valued literal. For instance, the minterms 100 and 010 merge into the literal 110. Thus, specifying a don't care set is not necessary because the don't care set is implicit in the multiple-valued algebra that allows the merging of different values into a single multiple-valued literal.

We further note that the minimized covers of Figures 3.4(b) and 3.5(b) are also very similar. In fact, they are identical except that the −'s in the 2nd column of Figure 3.4(b) are replaced with 1's in Figure 3.5(b). This is not surprising, since the merging of binary-valued implicants 100 and 010 under the don't cares 110 and 000 produces − − 0, whereas the merging of the three-valued minterms 100 and 010 produces the three-valued literal 110. A minimized symbolic cover is shown in Figure 3.5(c) that illustrates the symbolic values that have merged during minimization. There is a one-to-one correspondence between the multiple-valued literals of Figure 3.5(b) and the groups of symbolic values in Figure 3.5(c).

3.2.2   Input Constraints and Face Embedding

Face embedding constraints (input constraints or input relations) can be derived from the exactly minimized multiple-valued or symbolic cover such that if the symbolic variable is encoded satisfying these constraints then the cardinality of the resulting two-level Boolean implementation will be equal to the cardinality of the minimized multiple-valued cover.

A sufficient condition that ensures the preservation of the cardinality (of the multiple-valued cover minimized using ESPRESSO-MV)

inp1	00	10 110	01	super(00, 01)	= 0–	10 0– 01
inp2	01	11 011	01	super(01, 11)	= –1	11 –1 01
inp3	11	01 111	10	super(00, 01, 11)	= ––	01 –– 10

(a)   **(b)**   **(c)**

Figure 3.6: Satisfying constraints and cover construction

during the transition from the multiple-valued to the Boolean domain is to ensure that each multiple-valued input literal (group of symbolic values) in the minimized multiple-valued (symbolic) cover translates into a single cube in the Boolean domain. In other words, given a multiple-valued literal, the codes assigned to the symbolic values present in it should form a face in the Boolean n-space in such a way that the face does not include the codes of the symbolic values absent from the same multiple-valued literal. Such constraints are called **face-embedding** or **input** constraints.

We now summarize the salient features of the input encoding procedure.

- Given a face embedding constraint corresponding to a multiple-valued literal (group of symbolic values), the constraint is satisfied by an encoding if the super-cube of the codes of the symbolic values in the literal (group) does not intersect the codes of the symbolic values not in the literal (group).

- Satisfying all the constraints posed by a multiple-valued cover, guarantees that the encoded and minimized binary-valued cover will have a number of product terms no greater than the multiple-valued cover.

- Once an encoding satisfying all constraints has been found, a binary-valued encoded cover can be constructed directly from the multiple-valued (symbolic) cover, by replacing each multiple-valued literal (group of symbolic values) by the super-cube corresponding to that literal (group).

Codes satisfying the face-embedding constraints implied by the minimized multiple-valued cover of Figure 3.5(b) are shown in Figure 3.6(a). The super-cubes of each of the constraints are shown in

Figure 3.6(b), and they do not intersect the codes of the symbolic values not in the constraint. For instance, the super-cube $0- = enc(inp1)$ $\cup\ enc(inp3) = 00 \cup 01$ (corresponding to the multiple-valued literal 110) does not intersect $enc(inp3) = 10$ ($inp3$ is not in the constraint 110). Two binary encoding variables suffice to satisfy the face-embedding constraints (as compared to three in a one-hot code). Given that the constraints posed by the multiple-valued cover have been satisfied, we can directly construct a binary-valued encoded and minimized cover as shown in Figure 3.6(c). This represents an optimum solution to the input encoding problem, since the solution has minimum product term count, and uses a minimum number of encoding bits.

Note that the unused code 10 has been used as a don't care in the third implicant of Figure 3.6(c).

3.3   Satisfying Encoding Constraints

The preceding section described how encoding constraints could be generated for the input encoding problem. Given a set of constraints, one is required to find the minimum-length encoding for the symbols that satisfies these constraints. This is an NP-hard problem [83]. In this section, we describe heuristic algorithms for constraint satisfaction.

3.3.1   Definitions

The symbolic values in the symbolic input are denoted $V = \{v_0, v_1, \ldots, v_{n-1}\}$. We will represent the given set of input relations by an **input constraint matrix** A. The input constraint matrix A is a matrix $A \in \{0, 1\}^{r \times n}$

$$
\begin{bmatrix}
a_{0\ 0} & a_{0\ 1} & \cdots & a_{0\ n-1} \\
a_{1\ 0} & a_{1\ 1} & \cdots & a_{1\ n-1} \\
 & \cdots & \cdots & \\
a_{r-1\ 0} & a_{r-1\ 1} & \cdots & a_{r-1\ n-1}
\end{bmatrix}
$$

where r, the number of rows in the matrix equals the number of face or input constraints, and if $a_{ij} = 1$ then symbolic value v_j belongs to constraint i.

A row of the constraint matrix is said to be a **meet** if it represents the conjunction (bitwise AND) of two or more rows. A row of the constraint matrix is said to be **prime** if it is not a meet.

The **code matrix** S is a matrix $S \in \{0, 1\}^{n \times b}$

$$
\begin{bmatrix}
s_{0\ 0} & s_{0\ 1} & \cdots & s_{0\ b-1} \\
s_{1\ 0} & s_{1\ 1} & \cdots & s_{1\ b-1} \\
& \cdots & \cdots & \\
s_{n-1\ 0} & s_{n-1\ 1} & \cdots & s_{n-1\ b-1}
\end{bmatrix}
$$

where b is the number of encoding bits.

Our problem is to determine the code matrix S given the constraint matrix A. Note that for an implementable input encoding, the codes given to the symbolic values must be pairwise disjoint, *i.e.* each row of S must be pairwise disjoint.

3.3.2 Column-Based Constraint Satisfaction

In this section, we describe some basic properties of constraint matrices, and develop heuristic column-based methods that incrementally construct encodings which satisfy a given constraint matrix, one encoding bit at a time.

We first present some basic theorems that aid the search for a minimal-length encoding. These theorems were originally presented in [73].

Theorem 3.3.1 : *A one-hot coding corresponding to the identity code matrix $S = I \in \{0, 1\}^{n \times n}$ satisfies any constraint matrix A.*

Proof. Consider the ith row in A. Denote the set of symbolic values such that $a_{ij} = 1$, $Q1$, and the set of symbolic values such that $a_{ij} = 0$, $Q0$. We can obtain the super-cube corresponding to the ith row by bitwise OR'ing the codes of all the symbolic values in $Q1$. Since we have used a one-hot code, the super-cube c has 0's in all positions corresponding to the symbolic values not in $Q1$, *i.e.* those in $Q0$. Since the code of any symbolic value in $Q0$ has a 1 in exactly one of these positions where c has a 0, the super-cube c does not intersect the code of any symbolic value in $Q0$. □

Theorem 3.3.2 : *Given any constraint matrix A, $S = A^T$ satisfies the constraint matrix.*

Proof. Consider the ith row in A. Denote the set of symbolic values such that $a_{ij} = 1$, $Q1$, and the set of symbolic values such that $a_{ij} = 0$, $Q0$. Focus on the the ith column of S. For each of the symbolic values $v_k \in Q1$ we have $s_{ki} = 1$. For each of the symbolic values $v_k \in Q0$ we have $s_{ki} = 0$. Therefore the super-cube of the constraint corresponding to the ith row in A has a 1 in a position where all the codes given to the symbolic values not in the constraint have a 0. $\square$

Note that A^T may not be an implementable encoding. In this case, the codes of the symbolic values can be differentiated by adding an extra column (or columns) to S, namely $s \in \{0, 1\}^n$. The 0's and 1's in these extra columns can be arbitrary, as long as the addition results in all codes in S being pairwise disjoint.

Theorem 3.3.3 : *If S satisfies the constraint relation for a given A, then S' satisfies the constraint relation when S' is obtained from S by column permutation or column complementation.*

Proof. Column permutation in S obviously does not alter the satisfaction of the constraint matrix A. Complementing the ith column in S results in a change in the ith position of the super-cube $c = c_0, \dots c_{b-1}$ for any constraint $a \in A$. If $c_i = -$, it remains a $-$ after complementation of the codes. If $c_i = 0$ complementation results in $c_i = 1$, and vice versa. If the code of any particular symbolic value v_j not in a ($enc(v_j)$ $= s_{j0}, \dots, s_{j\,b-1}$) is such that $s_{ij} \cap c_i = \phi$, then complementation does not change this (because both s_{ij} and c_i are complemented).

     The above argument can be generalized to the case where multiple columns are complemented. $\square$

The following theorem allows for a compaction of the constraint matrix A.

Theorem 3.3.4 : *Consider the constraint matrix A' obtained from A by:*

1. *Removing all rows in A with a single 1.*

2. *Removing all rows in A with all 1's.*

3. *Removing all the meets of A.*

Then, if an implementable encoding S satisfies A' it satisfies A also.

Proof. Rows in A with a single 1 simply correspond to the constraint of an implementable encoding, *i.e.* that the code of each symbolic value be pairwise disjoint from each of the other symbolic values. Rows with all 1's do not correspond to any sort of encoding constraint.

Assume we have dropped a meet a from A. Without loss of generality assume that the meet is the conjunction of two rows in A', namely a^1 and a^2. Given the encoding S, the super-cubes corresponding to a^1 and a^2, namely c^1 and c^2, cover the super-cube corresponding to a, namely c. All the symbolic values that are not in a^1 or not in a^2 are not in a. The super-cube c does not intersect the codes of these values, since S satisfies a^1 and a^2, and c^1 and c^2 cover c. There are some symbolic values that are not in a but are in a^1 or a^2 (but not both). Consider one such value v_i. Assume this value v_i is in a^1. This implies that v_i is not in a^2, else it would be in a. If v_i is not in a^2, then c^2 does not intersect the code of v_i, since a^2 is satisfied by the encoding S. Since c_2 covers c, this implies that c does not intersect the code of v_i. Therefore, c does not intersect the codes of any symbolic values not in a, and S satisfies A.

The above argument can be generalized to the case where the meet is a conjunction of three or more rows. $\square$

Theorem 3.3.4 is used in a heuristic encoding strategy [71] to compact the constraint matrix A prior to finding an encoding that satisfies the compacted constraint matrix A'. The method used to find the encoding corresponds to selecting a row from A' and incrementally constructing the encoding as the transpose of the selected row.

Theorem 3.3.4 does not model all possible relationships between the rows of A and does not guarantee a minimum-length encoding. For instance consider the constraint matrix below:

$$\begin{bmatrix} 1 & 1 & 0 & 0 & 0 \\ 1 & 0 & 1 & 1 & 0 \\ 0 & 1 & 0 & 1 & 0 \\ 0 & 0 & 1 & 1 & 0 \end{bmatrix}$$

It cannot be compacted using Theorem 3.3.4. However, the encoding below obtained by dropping the last row in the constraint matrix and transposing the smaller constraint matrix, satisfies the original constraint matrix.

$$\begin{bmatrix} 1 & 1 & 0 \\ 1 & 0 & 1 \\ 0 & 1 & 0 \\ 0 & 1 & 1 \\ 0 & 0 & 0 \end{bmatrix}$$

The last row of the constraint matrix 00110 has a super-cube 01− under the above encoding which does not intersect any of the codes not in the constraint. The above example illustrates the limitations of Theorem 3.3.4. A method for column-based encoding to alleviate the above problem was presented in [31].

In the work of [31] a stronger compaction result regarding the constraint matrix A was presented.

Theorem 3.3.5 : *Given a constraint matrix, A, $S = A^T$ satisfies all constraints which are obtained by (bitwise) intersecting two or more rows in A or their (bitwise) complements in any combination.*

Proof. Construct a constraint, a, which is the bitwise intersection of the rows or complement of rows in A in any combination. We assume without loss of generality that a is constructed using all the rows in A in true or complemented form. (This is because if a is constructed using a subset of rows, then we can identify a matrix A^* corresponding to this subset of rows. We then have to prove that $S^* = A^{*^T}$ satisfies a, where a has been constructed using all the rows of A^*.) A bit vector b specifies if a has been constructed using the true or complemented form of each row in A. That is, $b[i] = 1$ if the ith row in A has been used in true form and $b[i] = 0$ if the ith row has been used in complemented form. Examining $S = A^T$, it is easy to see that the super-cube corresponding to the set of symbolic values in a is $c = b$. Each position in a that has a 1 corresponds to a column in A that equals b^T, and a row in S that equals b. The codes of symbolic values not in a differ from c in at least one bit position. If the code of a symbolic value, v_j, not in a, was the same as c then $b[i]$ would equal 1, and we have a contradiction. Since the codes of symbolic values not in a differ from c, S satisfies a. □

There is a subtle difference between the statements of Theorem 3.3.4 and Theorem 3.3.5. Theorem 3.3.4 corresponds to a weaker compaction procedure for the constraint matrix A, but states that any encoding that satisfies the compacted A' will satisfy A. Theorem 3.3.5 on the other hand, *requires* that the encoding that satisfies A and the compacted A' be constructed as the transpose of the compacted matrix A'.

Going back to our example that helped illustrate the limitations of Theorem 3.3.4, we see that the last row in the matrix shown below is the intersection of the complement of the first row with the true form of the second row, and can therefore be dropped from the matrix.

$$\begin{bmatrix} 1 & 1 & 0 & 0 & 0 \\ 1 & 0 & 1 & 1 & 0 \\ 0 & 1 & 0 & 1 & 0 \\ 0 & 0 & 1 & 1 & 0 \end{bmatrix}$$

An encoding algorithm can use Theorem 3.3.5 to compact the given constraint matrix. Given r constraints, there are $3^r - 1$ different ways of combining these as per Theorem 3.3.5. Thus, an exhaustive compaction is not possible for large examples. A constraint ordering approach is taken in [31], where the rows of the partially-compacted matrix A' are ordered such that the encoding corresponding to the transpose of the first k rows satisfies the entire matrix. Heuristics and simulated annealing can be used to find orderings that minimizes k.

The program NOVA [94] includes methods for finding minimum and minimal-length encodings satisfying a constraint matrix, as well as heuristic techniques to satisfy a maximal number of constraints, within a given number of encoding bits. The algorithms use a formulation of face embedding as one of detecting subgraph isomorphism.

3.3.3 Row-Based Constraint Satisfaction

The algorithm developed in [73] for satisfying constraints used a row-based approach, wherein codes were assigned to subsets of states at a time based on heuristic selection criteria. However, row-based methods have not proven effective in dealing with functions with a large number of symbolic values, and we will not describe these methods here.

3.3.4   Constraint Satisfaction Using Dichotomies

A **dichotomy** is a disjoint two-block partition of a subset of column indices of the constraint matrix A. The use of dichotomies for satisfying encoding constraints was suggested by Yang and Ciesielski [97] and Saldanha *et al* [83]. In this section, we describe methods for constraint satisfaction using dichotomies.

A **seed dichotomy** is associated with a row of A. A seed dichotomy associated with row i of A is a disjoint two-block partition $(l : r)$ such that the block l contains all the column indices j such that $a_{ij} = 1$, and r contains exactly one of column index k such that $a_{ik} = 0$. For instance consider the constraint matrix below:

$$\begin{bmatrix} 1 & 1 & 0 & 0 & 0 \\ 1 & 0 & 1 & 1 & 0 \\ 0 & 1 & 0 & 1 & 0 \\ 0 & 0 & 1 & 1 & 0 \end{bmatrix}$$

The seed dichotomies associated with the first row of this matrix are the following:

$$(0 \quad 1 \quad : \quad 2)$$
$$(0 \quad 1 \quad : \quad 3)$$
$$(0 \quad 1 \quad : \quad 4)$$

Each dichotomy is associated with a single bit in the encoding. This bit takes one value for all the symbols in the left block of the dichotomy, and takes the opposite value for all the symbols in the right block. Without loss of generality, let us assume that the symbols in the left block take the value 0 and the symbols in the right block take the value 1. Generating the minimum-length encoding involves finding the smallest number of dichotomies that cover all the seed dichotomies obtained from the reduced constrained matrix. This approach represents an exact algorithm for constraint satisfaction.

Two dichotomies are **compatible** if the two dichotomies do not require any symbol to have different values in an encoding bit. Therefore, two dichotomies d_1 and d_2 are compatible if the left block of d_1 is disjoint from the right block of d_2, and the right block of d_1 is disjoint from the left block of d_2. For example, the dichotomies

$$(0 \quad 1 \quad : \quad 2)$$
$$(1 \quad 3 \quad : \quad 2)$$
$$(0 \quad 1 \quad : \quad 4)$$

are compatible, while the dichotomies

$$(0 \quad 1 \quad : \quad 2)$$
$$(1 \quad 2 \quad : \quad 3)$$

are incompatible. Larger dichotomies are formed by taking the **union** of compatible dichotomies. The union of two dichotomies is formed by taking the union of the left blocks and the union of the right blocks of the two dichotomies. For example, the union of the compatible dichotomies

$$(0 \quad 1 \quad : \quad 2)$$
$$(1 \quad 3 \quad : \quad 4)$$

is the dichotomy

$$(0 \quad 1 \quad 3 \quad : \quad 2 \quad 4)$$

A dichotomy d_1 covers the dichotomy d_2 if the left and right blocks of d_2 are subsets of the left and right blocks, respectively of d_1, or are subsets of the right and left blocks, respectively of d_1. For example, in the dichotomies,

$$(0 \quad : \quad 1 \quad 2)$$
$$(0 \quad 3 \quad : \quad 1 \quad 2 \quad 4)$$
$$(1 \quad 2 \quad 3 \quad : \quad 0)$$
$$(0 \quad 1 \quad : \quad 2)$$

the first dichotomy is covered by the second and third dichotomies, but not by the fourth one. Based on this notion of covering, a **prime dichotomy** is defined as a dichotomy that is incompatible with all other dichotomies that it does not cover.

The exact constraint satisfaction algorithm in [83] is formulated as a prime generation and covering procedure as follows. In the first step, the dichotomies corresponding to the face-embedding constraints, and to the constraint that all symbols have unique codes are generated. In the second step, the prime dichotomies corresponding to these constraints are generated. The generation of primes is carried out by an approach similar to that of Marcus [67]. A literal is associated with each dichotomy, and each pair of incompatible dichotomies, d_1 and d_2 is written as the sum-term $(d_1 + d_2)$. The product of all such sums is obtained and re-written as an irredundant sum-of-products expression. For each cube in this irredundant sum-of-products expression, the dichotomies corresponding to the literals that are absent from the cube are identified. The union of each set of such dichotomies corresponds to a prime dichotomy. In order to multiply out the product-of-sums to a sum-of-products expression, the fact that the expression contains only 2-SAT terms is used to advantage to make the multiplication linear in the number of initial dichotomies. The final number of prime dichotomies might still be exponential. Once the primes are generated, unate covering (*cf.* Section 4.2.9) is used to identify the minimum cardinality set of primes that cover the initial set of dichotomies. The cardinality of this set is the minimum number of bits required to encode the symbols.

An example of the application of this procedure is shown in Figure 3.7. Figures 3.7(a) and (b) show the initial constraint matrix and the corresponding dichotomies, respectively. Figure 3.7(c) is the product-of-sums expression corresponding to the incompatible dichotomies, and Figure 3.7(d) is the associated irredundant sum-of-product expression. Figures 3.7(f) through (h) are the set of all primes, the minimum prime cover, and the minimum-bit encoding respectively.

The application of the exact algorithm is not feasible for any but the smallest sized FSMs. In practice, it is more relevant to attempt to satisfy as many constraints as possible within a fixed number of encoding bits. A heuristic algorithm for constraint satisfaction given the number of encoding bits was proposed in [83]. Briefly, the method is as follows. Initially, a disjoint partition on the symbols is obtained so that a minimal number of face-constraints have their symbols split across the partition. Each block in this partition is then recursively partitioned in the same manner. Each partition obtained in this manner represents a dichotomy of the symbols. Also note that if the blocks in a partition are P_1 and P_2, then the dichotomies obtained by recursively partitioning P_1 are compatible with the dichotomies obtained by recursively partitioning P_2. Therefore, the union of the sets of dichotomies obtained by recursively partitioning P_1 and P_2 can be performed to obtain larger dichotomies. Not all unions are performed. At any step, only those unions

	a – 12 : 3	
	b – 12 : 4	
	c – 12 : 5	
	d – 134 : 2	(a + d) (a + e) (a + f) (a + i) (a + j) (a + k)
	e – 134 : 5	(b + d) (b + e) (b + f) (b + g) (b + h) (b + i) (b + j) (b + k)
	f – 24 : 1	(c + d) (c + f) (c + i) (c + j)
	g – 24 : 3	(d + f) (d + g) (d + h) (d + i)
11000	h – 24 : 5	(e + f) (e + g) (e + i)
10110	i – 34 : 1	(f + j)
01010	j – 34 : 2	(g + i) (g + j) (g + k)
00110	k – 34 : 5	(h + j)
(a)	**(b)**	**(c)**

abcdefgh + abcdegj + abcdeijk + abcfghi + abdfgij + bdefijk + defghijk

(d)

12 : 3 4 5		
124 : 3 5		
1234 : 5		1 → 001
134 : 2 5		2 → 011
24 : 135	124 : 35	3 → 100
234 : 15	134 : 25	4 → 000
34 : 125	34 : 125	5 → 111
(f)	**(g)**	**(h)**

Figure 3.7: Example of exact satisfaction of input constraints using dichotomies

are performed that recover the face-embedding constraints whose symbols were split across the partition. If P_1 and P_2 correspond to the initial partition of symbols, then the union of the dichotomies obtained by recursively partitioning P_1 with the dichotomies obtained by recursively partitioning P_2 provides a set of prime dichotomies. From this set, an optimal set that satisfies a maximal number of constraints and which covers the seed dichotomies can be selected. It is claimed in [83] that a straightforward implementation of this algorithm leads to promising results.

3.3.5   Simulated Annealing for Constraint Satisfaction

Simulated annealing has been used to find a minimal-length encoding to satisfy a set of constraints, and to maximally satisfy a set of constraints using a fixed-length encoding [31, 60]. The constraint ordering approach of [31] lends itself to a simple and efficient simulated-annealing-based strategy to find minimal-length encodings that satisfy all the given constraints (*cf.* Section 3.3.2).

When attempting to satisfy a maximal number of constraints given a fixed encoding length two strategies are possible. In the first strategy, the constraint ordering formulation is used with the modification that the first b constraints, where b is the given fixed encoding length, are transposed to obtain an encoding. The total number of constraints that the encoding satisfies can be used as the cost function. However, merely using the above cost function does not produce good results, since it does not provide enough information to guide the random search, and further does not deterministically predict the number of product terms in an encoded and minimized result, if some constraints are left unsatisfied. Another problem with this approach is that transposing the first b constraints may not result in an implementable encoding.

In the second strategy, the symbolic values are assigned a random coding over the encoding length. Moves in the annealing step include an interchange of the codes given to two symbolic values, or the replacement of the code of a symbolic value by an unused code. The cost function can be computed in various ways, as we will elaborate in the sequel.

The implicants in the multiple-valued cover corresponding to the satisfied constraints map to single binary-valued implicants in the encoded and minimized cover. The unsatisfied constraints map to multiple cubes in the encoded and minimized cover. As an example consider the constraint 110 over the symbolic values $inp1$, $inp2$ and $inp3$, with $enc(inp1) = 01$, $enc(inp2) = 10$, and $enc(inp3) = 00$. This encoding does not satisfy the constraint 110, and any implicant that contains the multiple-valued literal 110, will be split into two cubes, one of which has 01 in its encoded binary-valued part, and the other has 10. It is necessary to estimate the number of cubes that a particular encoding will result in, for a constraint not satisfied by the encoding. Otherwise, we may satisfy a maximal number of constraints with a particular encoding, but the eventual minimized cover may be large due to each of the unsatisfied constraints resulting in a large number of cubes. The notion of partial satisfaction of constraints is a means of estimating [1] the number

---

[1]Note that the problem of exactly predicting the number of cubes is one of two-

of cubes that a given encoding produces for an unsatisfied constraint. A
distance-1 merge approach is used in the procedure of [31] to estimate
the number of cubes for each unsatisfied constraint. The total number of
cubes corresponding to the current encoding is computed incrementally
after each move of the annealing, and this cost is minimized.

A simpler method is to compute the number of symbolic values not in a given constraint whose codes intersect the super-cube corresponding to the constraint. The sum of these numbers over all the
constraints can be used as the cost function. For a satisfied constraint
this number is zero. The intuition is that if this number is large for a
constraint, then the current encoding will result in the constraint being
split up into a large number of cubes. Other cost functions have been
proposed as well in [60].

3.4   Input Encoding Targeting Multilevel Logic

In this section we will describe the work of Malik *et al* [56, 65]
in the encoding of functions with symbolic inputs targeting multilevel
implementations of logic. As described in Section 3.1, the input encoding problem targeting multilevel logic is significantly more difficult than
its two-level counterpart because of the very power of multilevel logic
transformations [14]. Approaches to state assignment (*e.g.* [26, 61])
that concentrate on finding common divisors that are cubes (or single product terms) are described in Section 5.4. In [56], techniques for
the multilevel optimization of logic with MV-input variables were developed. This permits the viewing of the optimizations largely independent
of the encoding, allowing the deferment of the encoding to the second
step. The main advantage of using this approach is that common divisors that are not restricted to single cubes can be considered. Common sub-expressions can be extracted from expressions with MV-inputs
while guaranteeing that this extraction and a subsequent encoding will
be sufficient to obtain all possible kernel intersections for all possible
encodings.

The notation used in this section has been defined in Section
2.2. In Section 3.4.1 the common sub-expression extraction technique,

---

level Boolean minimization.

originally published in [12], which is commonly used in multilevel logic optimization is reviewed. Notions of multiple-valued kernels are presented in Section 3.4.2. Section 3.4.3 describes the MV-factorization technique of [56]. The important issue of size estimation in algebraic multiple-valued factorization is the subject of Section 3.4.4. The actual encoding is done after multiple-valued factorization, and this step is described in Section 3.4.5.

3.4.1 Kernels and Kernel Intersections

We first review the process of common sub-expression extraction when there are no MV variables.

Common sub-expressions consisting of multiple cubes can be extracted from Boolean expressions using the algebraic techniques described in [12]. We review some definitions presented there.

Definition 3.4.1 *A **kernel**, k, of an expression g is a cube-free[2] quotient of g and a cube c. A **co-kernel** associated with a kernel is the cube divisor used in obtaining that kernel.*

As an example consider the expression $g = ae + be + z$ and the cube e. The quotient of g and this cube; g/e is $a + b$. No other cube is a factor of $a + b$, hence it is a kernel of g. The cube e is the co-kernel corresponding to this kernel.

An important result concerning kernels is that two expressions may have common sub-expressions of more than one cube if and only if there is a kernel intersection of more than one cube for these expressions [12]. Thus, we can detect all multiple-cube common sub-expressions by finding all multiple-cube kernel intersections. In [11] algorithms for detecting kernel intersections are described by defining them in terms of the rectangular covering problem. This approach was used in [56] to develop ideas in multiple-valued factorization for ease of understanding. However, the ideas hold with any technique for kernel extraction.

As an example of the rectangular covering formulation, con-

---

[2]No cube is an algebraic factor of k.

sider the expressions g_1 and g_2:

$$g_1 = \underset{1}{ak} + \underset{2}{bk} + \underset{3}{c} \quad ; \quad g_2 = \underset{4}{aj} + \underset{5}{bj} + \underset{6}{d}$$

The integer below a cube is a unique identifier for it. The kernels of g_1 and g_2 are shown below.

expression	co − kernel	kernel
g_1	1	$ak + bk + c$
g_1	k	$a + b$
g_2	1	$aj + bj + d$
g_2	j	$a + b$

The rectangular covering formulation for this example is shown below.

	a	b	ak	bk	aj	bj	c	d
1	0	0	1	2	0	0	3	0
k	1	2	0	0	0	0	0	0
1	0	0	0	0	4	5	0	6
j	4	5	0	0	0	0	0	0

The table above is a **co-kernel cube matrix** for the set of expressions. A row in the matrix corresponds to a kernel, whose co-kernel is the label for that row. Each column corresponds to a cube which is the label for that column. A non-zero entry in the matrix specifies the integer identifier of the cube (in the original expressions) represented by the entry.

A rectangle $\mathcal{R}$ is defined as a set of rows $S_r = \{r_0, r_1, \ldots, r_{m-1}\}$ and a set of columns $S_c = \{c_0, c_1, \ldots, c_{n-1}\}$ such that for each $r_i \in S_r$ and each $c_j \in S_c$, the (r_i, c_j) entry of the co-kernel cube matrix is non-zero. $\mathcal{R}$ covers each such entry. $\mathcal{R}$ is denoted as: $\{R(r_0, r_1, \ldots, r_{m-1}), C(c_0, c_1, \ldots, c_{n-1})\}$.

A rectangular covering of the matrix is defined as a set of rectangles that cover the non-zero integers in the matrix at least once (and do not cover a 0 entry). Once an integer is covered, all its other occurrences are replaced by don't cares (they may or may not be covered by other rectangles). Each rectangle that has more than one row indicates a kernel intersection. A covering for the above co-kernel cube matrix is: $\{R(2,4), C(1,2)\}$, $\{R(1), C(7)\}$, $\{R(3), C(8)\}$.

The kernel intersection $a+b$ between the two expressions is indicated by the first rectangle in the cover. The resulting implementation suggested by the covering is:

$$g_1 = kg_3 + c \quad ; \quad g_2 = jg_3 + d \quad ; \quad g_3 = a + b$$

We will use the total number of literals in the factored form of all the Boolean expressions as the metric for circuit size [10]. The above description has two fewer literals than the original description.

3.4.2   Kernels and Multiple-Valued Variables

Now consider the case where one of the input variables may be multiple-valued (MV). The following example has a single MV-variable X with six values and six binary-valued variables.

$$
\begin{aligned}
f_1 &= \quad X^{\{0,1\}}ak + \quad X^{\{2\}}bk + \quad c \\
&\qquad\quad 1 \qquad\qquad 2 \qquad\quad 3 \\
f_2 &= \quad X^{\{3,4\}}aj + \quad X^{\{5\}}bj + \quad d \\
&\qquad\quad 4 \qquad\qquad 5 \qquad\quad 6
\end{aligned}
$$

The integers below each cube are unique identifiers for that cube.

The definitions and matrix representations given in Section 3.4.1 are for binary-valued expressions of binary-valued variables. We now extend these to binary-valued expressions with MV-variables. As stated earlier, we consider only one of the variables to be MV, and the others are binary valued.

The definition of the kernel and co-kernel remain the same as in the binary-valued case. The definition of the **co-kernel cube matrix** is modified as follows:

Definition 3.4.2 *Each row represents a kernel (labeled by its co-kernel cube) and each column a cube (labeled by this cube). Each non-zero entry* *in the matrix now has two parts. The first part is the integer identifier of the cube in the expression (**cube part**). The second part is the MV-literal (MV-**part**) which when ANDed with the cube corresponding to its column and the co-kernel corresponding to its row forms this cube.*

The kernels of f_1 and f_2 are given below.

exp.	co-kernel	kernel
f_1	1	$akX^{\{0,1\}} + bkX^{\{2\}} + c$
f_1	k	$aX^{\{0,1\}} + bX^{\{2\}}$
f_2	1	$ajX^{\{3,4\}} + bjX^{\{5\}} + d$
f_2	j	$aX^{\{3,4\}} + bX^{\{5\}}$

The corresponding co-kernel cube matrix is given below:

	a	b	ak	bk	aj	bj	c	d
1	0	0	1	2	0	0	3	0
	0	0	$X^{\{0,1\}}$	$X^{\{2\}}$	0	0	1	0
k	1	2	0	0	0	0	0	0
	$X^{\{0,1\}}$	$X^{\{2\}}$	0	0	0	0	0	0
1	0	0	0	0	4	5	0	6
	0	0	0	0	$X^{\{3,4\}}$	$X^{\{5\}}$	0	1
j	4	5	0	0	0	0	0	0
	$X^{\{3,4\}}$	$X^{\{5\}}$	0	0	0	0	0	0

In the co-kernel cube matrix the cube part is given above the MV-part for each entry. The adjectives MV and binary will be used with co-kernel cube matrices and rectangles in order to distinguish between the multiple-valued and the binary case.

3.4.3   Multiple-Valued Factorization

A rectangle is defined as in the case for all binary variables with one modification. Now the rectangle is permitted to have zero entries also. Associated with each rectangle $\mathcal{R}$ is a **constraint matrix** $M^{\mathcal{R}}$ whose entries are the MV-parts of the entries of $\mathcal{R}$. For example, for the MV co-kernel cube matrix given above, M_1, is the constraint matrix for the rectangle $\{R(2,4), C(1,2)\}$.

$$M_1 = \begin{bmatrix} X^{\{0,1\}} & X^{\{2\}} \\ X^{\{3,4\}} & X^{\{5\}} \end{bmatrix}$$

A constraint matrix is said to be **satisfiable** if:

$$X^\alpha \subseteq \bigcup_j M(i,j) \bigwedge X^\alpha \subseteq \bigcup_i M(i,j) \Rightarrow X^\alpha \subseteq M(i,j)$$

i.e. if a particular value of X occurs somewhere in row i and also somewhere in column j, then it must occur in $M(i,j)$. M_1 given above can be shown to be satisfiable. If a constraint matrix is satisfiable then it can be used to determine a common factor between the expressions corresponding to its rows as follows. The union of the row entries for row i, ($\bigcup_j M(i,j)$) is ANDed with the co-kernel cube corresponding to row i. Similarly the union of the column entries for column j ($\bigcup_i M(i,j)$) is ANDed with the kernel cube corresponding to that column. Thus, this results in the following factorization of the expressions f_1 and f_2.

$$\begin{aligned} f_1 &= X^{\{0,1,2\}}k \ (X^{\{0,1,3,4\}}a + X^{\{2,5\}}b) + c \\ f_2 &= X^{\{3,4,5\}}j \ (X^{\{0,1,3,4\}}a + X^{\{2,5\}}b) + d \end{aligned}$$

Note that there is now a common factor between the two expressions which was not evident to start with. This common factor may be now implemented as a separate node in the MV-network and its output used to compute f_1 and f_2 as follows:

$$f_1 = X^{\{0,1,2\}}ky_3 + c \ ; \quad f2 = X^{\{3,4,5\}}jy_3 + d \ ;$$
$$f_3 = X^{\{0,1,3,4\}}a + X^{\{2,5\}}b$$

Not all matrices are satisfiable. An example of this is:

$$M_2 = \begin{bmatrix} X^{\{1\}} & X^{\{2\}} & X^{\{5\}} \\ X^{\{4\}} & X^{\{5\}} & X^{\{1\}} \end{bmatrix}$$

since $X^{\{5\}}$ occurs in row 1 as well in column 2 but is not present in $M(1,2)$.

A non-satisfiable matrix cannot be used to generate a common factor as shown above. This is because the addition to the row co-kernel cube ($X^{\{1,2,5\}}$ for row 1 in this example), and that to the column kernel cube ($X^{\{2,5\}}$ for column 2 for M_2) for the "offending" rows and column when intersected yield extra terms that are not present in $M(i,j)$. (The intersection is $X^{\{2,5\}}$ which contains $X^{\{5\}}$ which is not present in $M(1,2)$.)

Satisfiable constraint matrices are not the only source of common factors. In fact the condition can be relaxed as we see below.

Definition 3.4.3 M_r *is a* **reduced constraint matrix** *of M if* $\forall i,j$ $M_r(i,j) \subseteq M(i,j)$.

Note that this definition includes the original constraint matrix. An example of a reduced constraint matrix for M_1 is:

$$M_3 = \begin{bmatrix} X^{\{1\}} & X^{\{2\}} \\ X^{\{3,4\}} & X^{\{5\}} \end{bmatrix}$$

A reduced constraint matrix that is satisfiable can be used to generate a common factor by covering the remaining entries separately. For example with the original expressions and M_3 we can obtain the following factorization.

$$f_1 = X^{\{1,2\}}k(X^{\{1,3,4\}}a + X^{\{2,5\}}b) + X^{\{0\}}ak + c$$
$$f2 = X^{\{3,4,5\}}j(X^{\{1,3,4\}}a + X^{\{2,5\}}b) + d$$

Note that $X^{\{0\}}ak$ must be covered separately.

The notion of *incompletely-specified literals* is important in factorization of multiple-valued functions and is introduced through an example. Consider the following expressions in factored form:

$$\begin{aligned} f_7 &= X^{\{0,1,2\}} y_5 (X^{\{0,1,3,4\}} y_1 + X^{\{2,5\}} y_2) + y_3 \\ f_8 &= X^{\{3,4,5\}} y_6 (X^{\{0,1,3,4\}} y_1 + X^{\{2,5\}} y_2) + y_4 \\ f_9 &= X^{\{6\}} \end{aligned}$$

Note that f_7 could be modified as follows without changing the functional behavior.

$$f_7 = X^{\{0,1,2,6\}} y_5 (X^{\{0,1,3,4\}} y_1 + X^{\{2,5\}} y_2) + y_3$$

Replacing $X^{\{0,1,2\}}$ with $X^{\{0,1,2,6\}}$ leaves the function unchanged since the multiplicative factor, $(X^{\{0,1,3,4\}} y_1 + X^{\{2,5\}} y_2)$, does not have $X^{\{6\}}$ anywhere in it. Thus, $X^{\{0,1,2\}}$ may be *expanded* to $X^{\{0,1,2,6\}}$. With binary variables, expansion always results in a decrease in the number of literals in the circuit size since expanding a literal y_1 or $\bar{y}_1$ results in the removal of this literal. However, expansion of MV-literals does not result in their removal unless the expansion is to the literal 1. Thus, it is not clear if expansion is useful in decreasing circuit size. It depends on the encoding chosen.

Consider the encoded expressions of the unexpanded and expanded literals with the following encoding, E_1.

$$X^{\{0\}} : 000 \quad X^{\{1\}} : 100 \quad X^{\{2\}} : 001$$
$$X^{\{3\}} : 110 \quad X^{\{4\}} : 010 \quad X^{\{5\}} : 011$$
$$X^{\{6\}} : 101$$

$$\mathcal{E}(X^{\{0,1,2\}}, E_1) = \bar{s}_1 \bar{s}_2 + \bar{s}_1 \bar{s}_0$$
$$\mathcal{E}(X^{\{0,1,2,6\}}, E_1) = \bar{s}_1$$

With the following encoding, E_2, we get:

$$X^{\{0\}} : 000 \quad X^{\{1\}} : 100 \quad X^{\{2\}} : 001$$
$$X^{\{3\}} : 110 \quad X^{\{4\}} : 010 \quad X^{\{5\}} : 011$$
$$X^{\{6\}} : 111$$

$$\mathcal{E}(X^{\{0,1,2\}}, E_2) = \bar{s}_1$$
$$\mathcal{E}(X^{\{0,1,2,6\}}, E_2) = \bar{s}_1 + s_0 s_2$$

Thus, with E_1 expansion led to a smaller encoded implementation while with E_2 it led to a larger encoded implementation.

Since it is not possible to predict the effect of expansion until encoding, it is best if the decision to expand is deferred to the encoding step. This can be captured by permitting the MV-literal to be incompletely specified. An incompletely-specified MV-literal, $X^{S_1[S_2]}$, represents the incompletely-specified pseudo-function [3]:

$$X^{S_1[S_2]} = \begin{cases} 1 & \text{if } X \in S_1 \\ 0 & \text{if } X \notin S_1 \cup S_2 \end{cases}$$

---

[3]This is not a true function since the mapping is not uniquely defined for all elements of the domain.

The value of this pseudo-function is unspecified when $X \in S_2$. This permits f_7 to be expressed as:

$$f_7 = X^{\{0,1,2\}[6]} \, y_5 \, (X^{\{0,1,3,4\}} \, y_1 + X^{\{2,5\}} \, y_2) + y_3$$

This makes it convenient to express the fact the expansion is optional and the decision whether it should be done is left till the encoding step.

We are now in a position to see what the MV-factorization process gets us in terms of the final encoded circuit. We are interested in obtaining large common factors in the encoded implementation. The following theorem was proven in [56].

Theorem 3.4.1 *Let $\tilde{f}$ be a factor in $\tilde{\eta} = \mathcal{E}(\eta, E)$. Then there exists a factor f in η such that $\tilde{f} = \mathcal{E}(f, E)$.*

Proof. The proof follows from the fact that $\mathcal{E}$ is invertible. $\eta = \mathcal{E}^{-1}(\tilde{\eta}, E)$. Let, $f = \mathcal{E}^{-1}(\tilde{f}, E)$. Since $\tilde{f}$ is a factor in $\tilde{\eta}$, f must be a factor in η. $\square$

This result implies that if we consider two encodings, E_1 and E_2, then corresponding to factor $\tilde{f}_1$ for E_1, there is a factor $\tilde{f}_2$ for E_2. Thus there is a one-to-one correspondence between factors across different encodings. This implies that a choice of encoding can only determine the particular encoded form of the factor and not its existence. Thus, it may seem that we can always first select an encoding and then find the common factors later. However, the catch here is that the techniques for finding common factors in Boolean networks are algebraic and thus not strong enough to discover many good Boolean factors. Using conventional algebraic factorization after selecting an encoding exposes only a fraction of the common factors present. Thus, even though a common factor may exist in the Boolean network it would remain undetected. (It may also be useless in one network and valuable in the other. This is also related to the fact that for two functions to have a common factor the only condition necessary is that they have a non-zero intersection.) What is desirable is that we should at least be able to obtain factors that can be obtained using algebraic techniques and any possible encoding. The MV-factorization process, using reduced constraint matrices, aims to do exactly this. However, not all factors in all encoded implementations can be obtained by this process. The following example helps illustrate the limitations.

Consider the following expressions and encoding:

$$f_3 = X^{\{1,2,3\}} y_1$$
$$f_4 = X^{\{4,5,6\}} y_2$$

$$X^{\{1\}} : 101 \quad X^{\{2\}} : 111 \quad X^{\{3\}} : 110$$
$$X^{\{4\}} : 001 \quad X^{\{5\}} : 011 \quad X^{\{6\}} : 010$$

With s_0, s_1, s_2 as the encoding variables we obtain the following encoded implementation:

$$\tilde{f}_3 = y_1 \, s_0 \, s_1 + y_1 \, s_0 \, s_2$$
$$\tilde{f}_4 = y_2 \, \bar{s}_0 \, s_1 + y_2 \, \bar{s}_0 \, s_2$$

Here, $s_1 + s_2$ is a kernel common to both expressions. However it cannot be obtained by first doing MV-factorization and then selecting an encoding since no kernels exist for f_3 and f_4. In this case all the variables involved in this kernel intersection are in the encoding space. As a result the entire kernel intersection in the encoded implementation corresponds to a single MV-literal in the original circuit. However, any kernel intersection not comprised entirely of the encoding variables can be detected as the following theorem [56] states.

Theorem 3.4.2 *Let $\tilde{k}$ be a kernel intersection in $\tilde{\eta} = \mathcal{E}(\eta, E)$ not comprised entirely of encoding variables. Then there exists a common factor, k, extracted using the MV-factorization process for η (i.e. by using a reduced satisfiable constraint matrix) such that $\tilde{k} = \mathcal{E}(k, E)$.*

The above theorem shows that we are considering algebraic factors in the encoded implementations for all possible encodings. Not all algebraic factors consisting of only the encoding variables can be considered in the MV-network since they are strongly dependent on the encoding.

3.4.4 Size Estimation in Algebraic Decomposition

In algebraic factorization and decomposition of Boolean networks the sequence of operations plays a critical part in the quality of
the final results. The sequence referred to here is the sequence in which
the algebraic divisors (common kernels and cubes) are extracted. The
choice of a particular rectangle affects both the other available rectangles
as well as the cost of the rectangles. The exact (or globally optimum) solution to algebraic decomposition is an open problem [14]. The solution
accepted in the binary-valued case is a locally optimal one, *i.e.* a greedy
choice is made at each decision step. The evaluation of a particular
divisor is done by first estimating its size and then using this to determine the size reduction that would result if this divisor were selected.
Determining the size of a divisor is not a problem in the binary-valued
case since the size is directly measured in terms of the number of literals in the divisor. Consider the following example. The expression
$y_1 \, y_2 \, y_3 \, y_4 + y_1 \, y_2 \, y_3 \, y_5$ is factored as $y_1 \, y_2 \, y_3 \, (y_4 + y_5)$ by extracting
the common cube $y_1 \, y_2 \, y_3$. The size of the common cube is three literals
and extracting this common cube made it possible to represent the same
function in five literals instead of eight, a saving of three literals.

Unfortunately in the MV case there is no direct correspondence
between the size of the MV-literal and its final encoded implementation.
Consider the following example: $X^{\{0\}} \, y_1 \, y_2 + X^{\{1\}} \, y_1 \, y_3$ is factored as
$X^{\{0,1\}} \, y_1 \, (X^{\{0\}[2,3]} \, y_2 + X^{\{1\}[2,3]} \, y_3)$; X takes values from $\{0,1,2,3\}$.
Since the size of the encoded implementation of the MV-literals is a
function of the encoding selected, it is not possible to predict the value
of this factor in terms of the size reduction obtained in the encoded
implementation. In fact, depending on the encoding, the factor may
even result in an increase in size. Consider the following encoding E_1
and the encoded implementation of the initial and factored expressions.

$$X^{\{0\}} : 00 \quad X^{\{1\}} : 10$$
$$X^{\{2\}} : 01 \quad X^{\{3\}} : 11$$

Initial expression:

$$\bar{s}_0 \, \bar{s}_1 \, y_1 \, y_2 + s_0 \, \bar{s}_1 \, y_1 \, y_3$$

Factored expression:

$$\bar{s}_1 \, y_1 \, (\bar{s}_0 \, y_2 + s_0 \, y_3)$$

Thus, cube extraction results in a saving of two literals. However, consider the following encoding E_2.

$$X^{\{0\}} : 00 \quad X^{\{1\}} : 11$$
$$X^{\{2\}} : 01 \quad X^{\{3\}} : 10$$

Initial expression:

$$\bar{s}_0 \ \bar{s}_1 \ y_1 \ y_2 \ + \ s_0 \ s_1 \ y_1 \ y_3$$

Factored expression:

$$(\bar{s}_0\bar{s}_1 \ + \ s_0 \ s_1) \ y_1 \ (\bar{s}_0 \ y_2 \ + \ s_0 \ y_3)$$

In this case, MV-cube extraction actually results in an increase in size! Thus, the potential reduction in size obtained using a factor must be estimated taking into account the effect of encoding. However, no encoding exists yet; the minimization is being done in order to select one. This is a cyclic dependency which must be broken. This problem is faced in the expansion of incompletely-specified literals as well. Expanding a literal before encoding may or may not reduce the size after encoding.

3.4.5   The Encoding Step

The first attempt in [56] to model the effect of the encoding assumed that each MV-literal has an encoded implementation that is a single cube in the encoding space of minimum dimension, *i.e.* the encoding is minimum length[4]. This corresponds exactly to the two-level case, wherein the set of multiple-valued implicants in the two-level cover are encoded satisfying face-embedding constraints, such that each MV-implicant can be replaced by a single cube in the encoded cover. Unfortunately, satisfying all the face-embedding constraints typically requires substantially more bits than the minimum. Encodings close to the minimum length are desirable for area reasons, and when the length of the encoding is restricted, the above single-cube assumption turns out to be a poor estimate of the effect of the final encoding. In most cases the encoded implementation of each MV-literal will not be just a single cube. Thus there are limitations to this modeling, that was successful for two-level logic, in the multilevel case.

---

[4]The minimum number of bits needed to encode a p-valued variable is $\lceil \log(p) \rceil$.

This led to the next approach [56] in which an actual encoding is done for the current MV-network. However, the MV-literals are not replaced by their encoded implementations; the codes are just used in the size estimation for factoring. A minimum-length encoding is chosen using heuristics or simulated annealing, attempting to find a small total size for all the multiple-valued factors chosen. This approach has the advantage that the estimated size is valid for at least one encoded implementation. It has the disadvantage that a relatively slow step of encoding must be done for the estimation. The other potential problem with this is that the final encoding may be very different from the one selected during estimations. Experience with using this indicates that there are no big changes between the estimates and the results obtained with the final encoding. This is the approach currently being used in MIS-MV. However, there is potential for improvement here.

Experimental results indicate that MIS-MV outperforms other encoding programs that do not model common sub-expression extraction. Details of the implementation of MIS-MV can be found in [57].

3.5 Conclusion

A lot of theoretical groundwork has been laid for the input encoding problem in recent years. The fundamental notion of using symbolic minimization to predict the effects of the logic optimization *prior* to actually performing an encoding, has allowed the development of efficient encoding algorithms that target both two-level and multilevel logic.

There is room for improvement in algorithms that satisfy face-embedding constraints, attempting to minimize the number of encoding bits used. The notion of incompletely-specified literals introduced in Section 3.4.2 exists in the two-level case as well [71]. It may be possible to expand or reduce a multiple-valued implicant in a minimized multiple-valued cover. Expanding or reducing certain implicants may generate a set of face constraints that are easier to satisfy than the original set. Exploiting incompletely-specified literals can potentially improve the performance of two-level input encoding algorithms.

Using multiple-valued factorization [56] prior to encoding represents an elegant and satisfying analog to two-level multiple-valued minimization, when targeting multilevel logic. However, as indicated in Section 3.4.5, more theoretical and experimental work is required to accurately estimate the gains in extracting multiple-valued factors, enabling the selection of the best set of factors, prior to encoding.

Chapter 4

Encoding of Symbolic Outputs

The output encoding problem entails finding binary codes for symbolic outputs in a switching function so as to minimize the area or an estimate of the area of the encoded and optimized logic function. In this chapter, we will focus on describing output encoding algorithms that target two-level or Programmable Logic Array (PLA) implementations of logic. As in input encoding targeting two-level logic, the optimization step that follows encoding is one of two-level Boolean minimization.

We begin with an example. An arbitrary output encoding of the function shown in Figure 4.1(b), is shown in Figure 4.2(a). The

```
10 inp1    1010
01 inp1    0110
10 inp2    1010
-1 inp2    1011
1- inp3    0110
0- inp3    1001
-- inp4    0010
-- inp5    1101
```

```
0001    out1
00-0    out2
0011    out2
0100    out3
1000    out3
1011    out4
1111    out5
```

(a)   **(b)**

Figure 4.1: Symbolic covers

```
0001    001              0001    10000
00-0    010              00-0    01000
0011    010              0011    01000
0100    011              0100    00100
1000    011              1000    00100
1011    100              1011    00010
1111    101              1111    00001
```

(a)   **(b)**

Figure 4.2: Possible encodings of the symbolic output

symbolic values {*out*1,*out*2,out3,out4,*out*5} have been assigned the binary codes {001,010,011,100,101}. The encoded cover is now a multiple-output logic function. This function can be minimized using standard two-level logic minimization algorithms. These algorithms exploit the sharing between the different outputs so as to produce a minimum cover. It is easy to see that an encoding such as the one in Figure 4.2(b), where each symbolic value corresponds to a separate output, can have no sharing between the outputs. Optimizing the function of Figure 4.2(b) would produce a function with a number of product terms equal to the total number of product terms produced by *disjointly* minimizing each of the on-sets of the symbolic values of Figure 4.1(b). This cardinality is typically far from the minimum cardinality achievable via an encoding that maximally exploits sharing relationships.

In this chapter, we will first describe heuristic techniques that attempt to generate various types of *output constraints* which when satisfied result in maximal sharing during the two-level minimization step. De Micheli in [71] showed that exploiting dominance relationships between the codes assigned to different values of a symbolic output results in some reduction in the overall cover cardinality. This work in described in Section 4.1. Disjunctive relations were shown to be an additional sharing mechanism by Devadas and Newton in [30], and the notion of generalized prime implicants (GPIs) was introduced. GPIs provide a systematic means to exactly minimize the product-term count in the encoded, two-level implementation, using a strategy similar to the Quine-McCluskey [68, 78] method. This exact solution to output encoding is described in Section 4.2. Heuristic strategies based on this method have been developed and are also described in Section 4.2.

We assume we are given a symbolic cover S with a single symbolic output. [1] The different symbolic values are denoted v_0, .., v_{n-1}. The encoding of a symbolic value v_i is denoted $enc(v_i)$. The on-sets of the v_i are denoted ON_i. Each ON_i is a set of D_i minterms $\{m_{i1}, .. m_{iD_i}\}$. Each minterm m_{ij} has a tag as to what symbolic value's on-set it belongs to. Note that a minterm can only belong to a single symbolic value's on-set. Minterms are also called 0-cubes in this section.

4.1   Heuristic Output Encoding Targeting Two-Level Logic

We describe heuristic output encoding methods in this section that exploit dominance relations. We illustrate the limitations of these methods, by presenting the notion of disjunctive relations.

4.1.1   Dominance Relations

Some heuristic approaches to solving the output encoding problem have been taken in the past (*e.g.* [71, 82]). The program CAPPUC-CINO [71] attempts to minimize the number of product terms in a PLA implementation and secondarily the number of encoding bits.

The output encoding algorithm in CAPPUCCINO is based on exploiting *dominance* relationships between the binary codes assigned to different values of a symbolic output. For instance, in the example of Figure 4.1(b), if the symbolic value *out1* is given a binary code 110 which dominates the binary code 100 assigned to *out2*, then the input cubes corresponding to *out1* can be used as don't cares for minimizing the input cubes of *out2*. Using these don't cares can reduce the cardinality of the on-set of the symbolic value *out2*.

This is illustrated in Figure 4.3. In Figure 4.3(a) the first three implicants of the symbolic cover of Figure 4.1(b) are shown. In Figure

---

[1]Multiple symbolic outputs may be present in the given function. The exact output encoding algorithm presented in this chapter is generalized to the multiple symbolic case in Section 4.2.6. The other algorithms presented here can be generalized as well.

```
0001   out1        0001   110        0001   110
00-0   out2        00-0   010        00--   010
0011   out2        0011   010
```

(a)   **(b)**   **(c)**

Figure 4.3: Illustrating dominance relations

4.3(b), *out1* has been coded with 110 and *out2* with 100. The encoded
cover has been minimized to a two-cube cover in Figure 4.3(c). Note
that the input minterm 0001 of *out1* has been merged into the single
cube 00 − − that asserts the code of *out2*. Note that 00 − − contains
the minterm 0001 that asserts *out1*.

To summarize, if $enc(v_i) \succ enc(v_j)$, then ON_i can be used as a
DC-set when minimizing ON_j. This is the basis of the output encoding
strategy described in the next section.

4.1.2 Output Encoding by the Derivation of Dominance Relations

In CAPPUCCINO, dominance relationships between symbolic values that result in maximal reduction of the on-sets of the dominated
symbolic values are heuristically constructed. Satisfying these dominance relationships (which should not conflict) results in some reduction
of the overall cover cardinality. However, minimum cardinality cannot
be guaranteed because all possible dominance relations are not explored,
nor is an optimum set selected.

A dominance graph G_D corresponds to a graph where each
node is a symbolic value, and a directed edge from v_i to v_j implies that
$v_i \succ v_j$. The following theorem is proven in [71].

Theorem 4.1.1 : *A set of dominance relations represented by a dominance graph G_D can be satisfied if and only if the dominance graph does
not contain any cycles.*

Proof. The only if part is obvious. If G_D does not have any cycles, we
can levelize the graph, assigning level 0 to the nodes that do not have
any directed edges fanning out of them, level 1 to the nodes that only

dominance-encode():

```
{
    P = φ ;
    Begin with no edges in G_D ;
    for ( k = 0; k < n; k = k + 1) {
        i = select ( k ) ;
        J = { j | ∃ a path from v_i to v_j in G_D } ;
        OFF_i = ∪_J ON_j ;
        DC_i = ON_i ∪ OFF_i ;
        M_i = minimize ( ON_i, OFF_i, DC_i ) ;
        foreach ( j ∉ J && j ≠ i ) {
            if ( M_i ∩ ON_j ≠ φ )
                add edge from v_j to v_i in G_D ;
        }
        P = P ∪ M_i ;
    }
}
```

Figure 4.4: Procedure for symbolic minimization in CAPPUC-CINO

fan out to level 0 nodes, and so on. We choose our encoding length to be n, *i.e.* the number of symbolic values. We arbitrarily order the nodes at each level. We encode the bottom-most level 0 node with the n-bit code that has a 1 in the first position, and 0's in all other positions. The next node is coded with the code that has 1's in the first two positions, and 0's in the remaining positions, and so on. □

The symbolic minimization loop of CAPPUCCINO is illustrated in Figure 4.4.

To minimize ON_i, an explicit representation of the corresponding don't care set is computed in the loop. Given a current dominance graph G_D, the off-set for v_i is defined as the set of product terms corresponding to the on-sets of all symbolic values v_j dominated by v_i. That is, a path exists from v_i to v_j in G_D. The DC-set for v_i, namely DC_i is computed by complementing the union of ON_i and OFF_i. Given the on-set, off-set and DC-set for the symbolic value v_i, a two-level minimizer is invoked to obtain a minimized representation M_i. All the symbolic values $v_j \neq v_i$ which are not dominated by v_i are checked to see if terms in their on-sets have been used in M_i. If any term from ON_j has been used in M_i, and if there does not already exist an edge from v_j to v_i in G_D, a new edge is added from v_j to v_i in G_D.

The procedure **select**() sorts the symbolic values according to a heuristic criterion. Procedure **minimize**() is a call to a two-level binary-valued output minimizer like ESPRESSO. The procedure **dominance-encode**() generates a set of minimized on-sets for each of the symbolic values in P, and a acyclic dominance graph G_D. The following theorem can be proven about the above procedure, provided an exact two-level minimization is used both in **minimize**() and after output encoding.

Theorem 4.1.2 : *An encoding of symbolic values that satisfies all the dominance relations in G_D, will result in an encoded and minimized cover P^E such that $||P^E|| \leq ||P|| \leq ||P^O||$, where P corresponds to the cover constructed by the procedure and P^O corresponds to the cover obtained by one-hot coding the symbolic values.*

Proof. We merely give a sketch of the proof. P has been constructed using only the dominance relations in G_D. The encoding chosen is guaranteed to satisfy all the dominance relations in G_D, and perhaps others as well. This implies that during exact multiple-output Boolean minimization, the don't care sets due to dominance relations in the encoded function are at least as large as the don't care sets corresponding to the construction of P. Therefore $||P^E|| \leq ||P||$. $||P|| \leq ||P^O||$ because the one-hot code does not satisfy any dominance relations. $\square$

Note that the cardinality of P^E and P may be far from optimum. One reason for this is because of the heuristics used in **select**(), and the possibility of a suboptimal set of dominance relations being chosen in G_D. However, a more basic shortcoming of CAPPUCCINO is that dominance relations are not the only kind of relationships between symbolic values that can be exploited, in symbolic output minimization. As will become clear in Section 4.1.4, after a symbolic cover has been encoded, it represents a multiple-output logic function and minimizing

```
00011    out1 1
00000    out1 1
10100    out1 1
01100    out1 1
00010    out2 1
00001    out2 1
11001    out2 1
00100    out3 1
1110-    out3 1
01001    out4 1
10001    out4 1
```

Figure 4.5: Example function to illustrate dominance-relation-based encoding

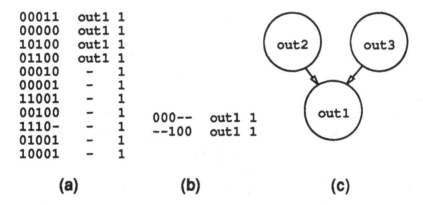

(a)　　　　　　　**(b)**　　　　　　　**(c)**

Figure 4.6: First step of constructing dominance relations

a multiple-output function entails exploiting other sharing relationships other than just dominance.

We give an example that illustrates the procedure **dominance-encode()**. In Figure 4.5, we have a function with a symbolic output that can take on four possible values, and a single binary-valued output. Each of the on-sets of the symbolic values have been disjointly minimized. In Figures 4.6 through 4.8, the various steps in the procedure are illustrated. We first pick $out1$ as the symbolic value whose on-set we wish to minimize. In Figure 4.6(a) the on-set of the symbolic value $out1$, ON_1 is minimized with ON_2, ON_3 and ON_4 as don't cares. This results in a reduction of cover cardinality for ON_1 from four to two, and

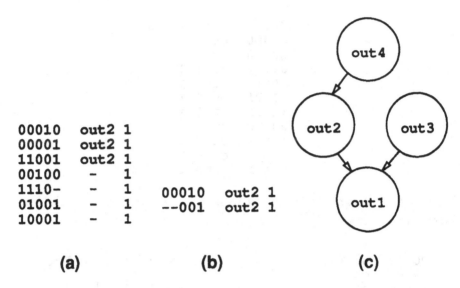

```
00010   out2  1
00001   out2  1
11001   out2  1
00100    -    1
1110-    -    1        00010   out2  1
01001    -    1        --001   out2  1
10001    -    1
```

(a)   **(b)**   **(c)**

Figure 4.7: Second step of constructing dominance relations

the minimized M_1 shown in Figure 4.6(b). We note that the ON_2 and ON_3 intersect the minimized M_1. Therefore we add two edges to the dominance graph G_D, as shown in Figure 4.6(c). We next pick *out2* and minimize ON_2 with *out3* and *out4* as don't cares as shown in Figure 4.7(a). The minimized M_2 is shown in Figure 4.7(b). M_2 has one less product term than ON_2. Only ON_4 intersects M_2 therefore we add a single edge to the dominance graph as shown in Figure 4.7(c).

We can now encode the four symbolic values using the encoding shown in Figure 4.8(a) so as to satisfy the dominance graph constructed by the procedure (shown in Figure 4.7(c)). Substituting the encoding into the function of Figure 4.5(a) and minimizing the binary-valued function gives us the function of Figure 4.8(b). A gain of three product terms over a disjointly minimized implementation has been achieved.

4.1.3 Heuristics to Minimize the Number of Encoding Bits

We will postpone the description of heuristics that can be used to minimize the number of encoding bits required to satisfy a set of dominance relations to Section 5.1.3 of Chapter 5.

```
                              --100   00 1
          out1 ➤ 00          000--   00 1
          out2 ➤ 01          --001   01 1
          out3 ➤ 10          00010   01 0
          out4 ➤ 11          00100   10 0
                             10001   10 0
                             01001   10 0
                             1110-   10 1
```

(a)   **(b)**

Figure 4.8: Encoding satisfying dominance relations

4.1.4 Disjunctive Relationships

Consider the symbolic cover of Figure 4.9(a). The function has one symbolic output and one binary-valued output. Using dominance relationships alone in an encoding, it is not possible to reduce the size of any of the on-sets of the symbolic values. One such encoding is shown in Figure 4.9(b), with *out1* given the binary code 00, *out2* given 01 and *out3* given 11. However, if *out1* were coded with 11, *out2* with 01 and *out3* with 10 as in Figure 3(c), one obtains a reduction in cover cardinality after minimization (Figure 4.9(d)). Note that in a dominance relationship, the on-set of the *dominated* symbolic value is reduced. However, in Figures 4.9(c) and 4.9(d), it is in fact the *dominating* symbolic value, *out1*, whose on-set cardinality has been reduced from 1 to 0. This is because of the *disjunctive relationship* between the codes of *out2*, *out3* and *out1*. *out1* = *out2* | *out3* and hence the on-set of *out1* can be reduced using the on-set of *out2* and *out3*. Just making *out1* dominate *out2* and *out3* is not enough, the code of *out1* has to be the disjunction (bitwise OR) of the codes of *out2* and *out3*. Exploiting these relationships is basic to a multiple-output logic minimizer and an exact encoding algorithm has to take into account these relationships in order to produce a minimum cardinality cover after optimization. Disjunctive relations may involve any number of symbolic values. For instance, the code of a symbolic value may be the bitwise OR of three other symbolic value codes.

```
101   out1 1              101   00 1
100   out2 1              100   01 1
111   out3 1              111   11 1
      (a)                       (b)


101   11 1
100   01 1               10-   01 1
111   10 1               1-1   10 1
      (c)                       (d)
```

Figure 4.9: Dominance and disjunctive relationships

4.1.5 Summary

The work of [71] implemented in the program CAPPUCCINO represented the first systematic effort to heuristically solve the output encoding problem targeting two-level logic. The methods used were generalized to be applicable to the state assignment problem as well, and we will describe this extension in Section 5.1.2.

One could envision using a CAPPUCCINO-style strategy that takes into account disjunctive as well as dominance relations in order to arrive at an exact output encoding algorithm. However, enumerating dominance or disjunctive relationships is very time-consuming. Finding the reduction in cover cardinality that can be accrued via an encoding satisfying each dominance or disjunctive relationship requires an exact logic minimization. Further, these relationships interact in complex ways and their effects are not simply cumulative. To solve the output encoding problem systematically and exactly, one has to modify the prime implicant generation and covering strategies that are basic to two-level Boolean minimization. We describe such a method in the next section.

4.2 Exact Output Encoding Targeting Two-Level Logic

In this section, an exact algorithm for output encoding, originally published in [30], that guarantees a minimum cardinality encoded cover is presented.

The algorithm finds an encoding that minimizes the number of product terms in an optimized PLA implementation. The algorithm consists of the following steps:

1. Generation of generalized prime implicants (GPIs) from the original symbolic cover.

2. Solution of a constrained covering problem involving the selection of a minimum number of GPIs that form an encodeable cover.

3. Encoding of the symbolic outputs respecting the encoding constraints generated during Step 2.

4. Given the codes of the symbolic outputs and the selected GPIs, a PLA with product term cardinality equal to the number of GPIs can be trivially constructed. This PLA represents an exact solution to the encoding problem.

Various techniques to generate GPIs that are modifications to classical prime implicant generation techniques can be used in Step 1. The covering problem of Step 2 is more complex than the unate covering problem and hence classical covering algorithms cannot be directly used. Step 3 involves constrained encoding where the objective is to minimize the number of encoding bits required to satisfy the constraints. This step is NP-hard. The focus here is to exactly minimize PLA product term cardinality and heuristically minimize PLA area.

These steps involved in the generation of generalized prime implicants (GPIs) and the definition of the encodeability of a set of GPIs are described in detail in Sections 4.2.1 through 4.2.7. In Section 4.2.8, it is described how various techniques for generating the prime implicants of binary-valued output functions can be used to generate all the GPIs for functions with symbolic outputs. In Section 4.2.9, strategies are reviewed for solving the classical covering problem and an approach to solving the covering problem with associated encodeability constraints is described.

```
1101   out1        1101   (out1)
1100   out2        1100   (out2)      110-   (out1,  out2)
1111   out3        1111   (out3)      11-1   (out1,  out3)
0000   out4        0000   (out4)      000-   (out4)
0001   out4        0001   (out4)
```

     (a)   **(b)**

Figure 4.10: Generation of generalized prime implicants

4.2.1 Generation of Generalized Prime Implicants

The generation of generalized prime implicants (GPIs) proceeds as in the well-known Quine-McCluskey (Q-M) procedure [68], with some differences.

1-cubes are constructed by merging all pairs of mergeable 0-cubes. Recall that each minterm m_{ij} has a tag as to what symbolic value's on-set it belongs to. If two 0-cubes with the same tag, (v_i), are merged then the 1-cube has the same tag (v_i). On the other hand, if a 0-cube of tag (v_i) is merged with a 0-cube with tag (v_j), the resultant 1-cube has a tag (v_i, v_j). The rule for canceling 0-cubes covered by 1-cubes is also different from the Q-M method. A 0-cube can be canceled by a 1-cube if and only if their tags are identical. A 1-cube 11— with tag (v_1, v_2) cannot cancel a 0-cube 110 with tag (v_1).

The above can be generalized to the k-cube case.

1. When two k-cubes merge to form a $k + 1$-cube, the tag of the $k + 1$-cube is the *union* of the two k-cube tags.

2. A $k + 1$-cube can cancel a k-cube only if the $k + 1$-cube covers the k-cube and they have identical tags.

A cube with a tag that contains all the symbolic values $(v_0, \ldots v_{n-1})$ can be discarded and is not a GPI. These cubes are not required in a minimum solution (*cf.* Theorem 4.2.1). The generation of generalized prime implicants for the symbolic cover of Figure 3(a) is shown in Figure 4.10. There are 6 GPIs with associated tags.

4.2.2 Selecting a Minimum Encodeable Cover

Given all the GPIs, one has to select a minimum subset of GPIs such that they cover all the minterms and form an *encodeable* cover. If the additional restriction of encodeability for a selected subset of GPIs was absent, then the output encoding problem would be equivalent to two-level Boolean minimization. The selection is carried out by solving a covering problem (Section 4.2.9 deals with the covering problem). In the sequel, the meaning of an encodeable cover is described.

Consider a minterm, m, in the original symbolic cover S. Let the minterm belong to the on-set of v_m. Obviously, in any encoded cover the minterm m has to assert the code given to v_m, namely $enc(\ v_m\ )$. Let the selected subset of GPIs be $p_1,\ ..,\ p_G$. Let the GPIs that cover m in this selected subset be $p_{m,1},\ ..,\ p_{m,M}$. For functionality to be maintained, the following relation has to be satisfied, for all minterms $m \in S$.

$$\bigcup_{i=1}^{M} \bigcap_{j} enc(\ v_{p_{m,i},\ j}\ )\ =\ enc(\ v_m\ )\ \ \forall\ m \tag{4.1}$$

where the $v_{p_{m,i},\ j}$ represent the symbolic values that are in the tag of the GPI $p_{m,i}$. In Figure 4.11 is shown a selection of GPIs for the symbolic cover of Figure 4.10(a) (whose GPIs are enumerated in Figure 4.10(b)). The GPIs $110-$, $11-1$ and $000-$ have been selected from Figure 4.10(b) in Figure 4.11(a). The constraints corresponding to Eqn. 4.1 for each minterm are given in Figure 4.11(b). The minterm 1101 is covered by both selected GPIs, one of which has a tag ($out1$, $out2$) and the other has a tag ($out1$, $out3$). Therefore, Eqn. 4.1 specifies:

$$enc(out1)\ \cap\ enc(out2)\ \bigcup\ enc(out1)\ \cap\ enc(out3)\ =\ enc(out1)$$

for the minterm 1101 and other constraints for the remaining minterms. If a minterm is covered by a GPI with the same tag as the minterm, then the constraint specified by the minterm via Eqn. 4.1 is an identity.

Eqn. 4.1 gives a set of constraints on the codes of the symbolic values, given a selection of GPIs. If an encoding can be found that satisfies all these constraints, then the selection of GPIs is encodeable. However, a selection of GPIs may have an associated set of constraints that are mutually conflicting.

```
110-    (out1,  out2)
11-1    (out1,  out3)
000-    (out4)
```

(a)

```
1101    enc(out1) ∩ enc(out2) ∪ enc(out1) ∩ enc(out3) = enc(out1)
1100    enc(out1) ∩ enc(out2) = enc(out2)
1111    enc(out1) ∩ enc(out3) = enc(out3)
0000    enc(out4) = enc(out4)
0001    enc(out4) = enc(out4)
```

(b)

Figure 4.11: Encodeability of selected GPIs

4.2.3   Dominance and Disjunctive Relationships to Satisfy Constraints

We now make the interesting observation that the constraints specified by Eqn. 4.1 can be satisfied by means of dominance and disjunctive relations between symbolic values. Continuing with our example, to satisfy

$$enc(out1) \cap enc(out2) \bigcup enc(out1) \cap enc(out3) = enc(out1)$$

one has three alternatives:

1. $enc(out2) \succ enc(out1)$

2. $enc(out3) \succ enc(out1)$

3. $enc(out1) \preceq enc(out2) \mid enc(out3)$

Given an arbitrary constraint, a set of dominance and disjunctive relationships can be derived such that satisfying any single relation satisfies the constraint. Dominance and disjunctive relationships may conflict across a set of constraints. For instance, one cannot satisfy both $enc(out1) \succ enc(out2)$ and $enc(out2) \succ enc(out1)$. This represents a cyclic set of constraints. Also, if one picks the equality in choice (3) above, then it is required that $enc(out1) \succ enc(out2)$ and $enc(out1) \succ enc(out3)$. In that case, one cannot satisfy both (1) and (3) with the same encoding.

Given a selection of GPIs, a set of constraints is derived via Eqn. 4.1 and a graph is constructed where each node represents a symbolic value. Directed edges in the graph represent dominance relations and undirected edges enclosed by arcs represent disjunctive relations. Each directed edge and arc has a label, corresponding to the minterm that produces the constraint represented by the edge or arc. The graph corresponding to the selected GPIs of Figure 4.11 is shown in Figure 4.12(a). A directed edge from *out*1 to *out*2 implies the code of *out*1 should dominate the code of *out*2. The dotted arc around the two undirected edges emanating from *out*1 implies that the code of *out*1 should be equal to or be dominated by the disjunction (bitwise OR) of the codes of its fanout symbolic values, in this case, *out*2 and *out*3. That is, $enc(out1) \preceq enc(out2) \mid enc(out3)$. *out*1 is called the **parent** in the disjunctive arc and *out*2 and *out*3, the **siblings** in the disjunctive arc. The disjunctive arc specifies equality or dominance, however, due to other relationships equality may be specifically required. In the case of disjunctive dominance the edges will be undirected, in the case of disjunctive equality the edges will directed towards the siblings to indicate that the parent dominates the siblings.

The graph corresponding to a selection of GPIs is encodeable and logic functionality is maintained, if four conditions are met. One selects either an edge or an arc of each label. In the case of selecting an arc, all dominance edges covered by the arc (implied by the disjunctive relationship) are also selected. For some selection,

1. There should be no directed cycles in the graph.

2. The siblings in any disjunctive arc should not have directed paths between each other.

3. No two disjunctive equality arcs can have exactly the same siblings and different parents.

4. The parent of a disjunctive dominance (equality) arc should not dominate (any symbolic value/node that dominates) all the siblings in the arc.

The graph of Figure 4.12(b), derived from the graph of Figure 4.12(a), satisfies these properties and hence the selection of GPIs is

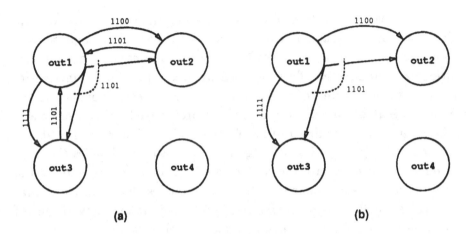

Figure 4.12: Encodeability graphs

valid. This implies that one can find an encoding such that the optimized cover has 2 product terms.

Given a constraint specified by Eqn. 4.1 of the form

$$a \cap b \cap c \; \bigcup \; a \cap d \cap e \; \bigcup \; a \cap f \cap g \; = \; a \qquad (4.2)$$

one has more complex choices than the equation in our example. To satisfy $a \cap b \cap c = a$, for instance, it is required to satisfy both $b \succ a$ and $c \succ a$. This merely corresponds to a pair of directed edges that have to be selected simultaneously. Further, one can satisfy $a \cap b \cap c \; \bigcup \; a \cap d \cap f = a$ by satisfying $b \cap c \; \bigcup \; d \cap f \; \succeq \; a$. This corresponds to a disjunctive relationship with nested conjunctive terms. The siblings here are **conjunctive nodes** $b \cap c$ and $d \cap f$. These conjunctive nodes are dominated by b, c and d, f respectively. Conditions 2-4 should be satisfied for arcs whose siblings are conjunctive nodes as well. The symbolic values whose conjunction forms the conjunctive node are called the **immediate ancestors** of the node. The symbolic values whose codes are to dominate a given symbolic value are called the **ancestors** of the symbolic value. The immediate ancestors dominate the conjunctive node. Also, if all the immediate ancestors dominate a particular symbolic value, then the conjunctive node also dominates that value. For instance, if all the immediate ancestors of a conjunctive node dominate the parent of a disjunctive arc that the node is a sibling of, then there exists a cycle in the graph rendering it unencodeable.

110-	(out1, out2)
11-1	(out1, out3)
000-	(out4)

out1	→ 11
out2	→ 01
out3	→ 10
out4	→ 00

110-	01 1
11-1	10 1
000-	00 1

(a) **(b)** **(c)**

Figure 4.13: Constructing the optimized cover

4.2.4 Constructing the Optimized Cover

If a selection of GPIs has been made that covers all minterms and is encodeable, then an encoding can be trivially found that satisfies the constraints (see Theorem 4.2.2). It is now possible to construct an encoded and optimized cover. The cover will contain the selected GPIs. For each GPI, the output combination in the cover is found using the tag corresponding to the GPI. The codes corresponding to all the symbolic values in the tag of the GPI are intersected (bitwise ANDed) to produce the output part. Continuing with our example, the GPIs selected and the tags for the GPIs are shown in Figure 4.13(a). These GPIs have an associated graph that is encodeable and an encoding satisfying the constraints is given in Figure 4.13(b). Note that the encoding has to satisfy disjunctive equivalence $enc(out1) = enc(out2) \mid enc(out3)$, rather than disjunctive dominance $enc(out1) \prec enc(out2) \mid enc(out3)$. This is because of the dominance relationships $enc(out1) \succ enc(out2)$ and $enc(out1) \succ enc(out3)$. The optimized cover is constructed with the GPIs by intersecting the codes of symbolic values in the tags of each GPI to obtain the output part (Figure 4.13(c)).

4.2.5 Correctness of the Procedure

Proposition 4.2.1 *The selection of a minimum cardinality encodeable cover from the GPIs represents an exact solution to the output encoding problem.*

The remainder of this section is devoted to justifying Proposition 4.2.1. It is shown first that logic functionality is retained.

Lemma 4.2.1 *Satisfying Eqn. 4.1 and constructing the output part as shown in the previous section retains logic functionality.*

Proof. The output part of a GPI is constructed by intersecting all the codes of the symbolic values contained in its tag. That is precisely the intersection term in Eqn. 4.1. The output of a minterm in a PLA is the OR of all the outputs asserted by the cubes that cover the minterm. This corresponds to the union (OR) in Eqn. 4.1. Thus, satisfying Eqn. 4.1 implies that each minterm asserts the same output combination as it would have in the original encoded but unoptimized cover. □

Next, it is shown that the canceled k-cubes during GPI generation are not necessary in a minimum solution.

Theorem 4.2.1 *A minimum cardinality encodeable solution can be made up entirely of GPIs.*

Proof. Assume that there is a minimum cardinality solution with a cube c_1 that is not a GPI. Let the tag of c_1 be T. We know that a GPI p_1 exists such that $p_1 \succ c_1$ and such that the tag of p_1 is T. Replacing c_1 with p_1 will not change the cardinality of the cover. The minterms corresponding to $p_1 - c_1$ will be covered by an extra GPI p_1 and therefore Eqn. 4.1 for those minterms will be different. However, the extra tag in the equation merely represents an extra option in the graph corresponding to the encodeability. Since the original graph was encodeable, adding edges with the same label as the labels of edges originally contained in the graph will not change the encodeability.

Also discarded are cubes with tags that contain all the symbolic values. If such a cube exists in a minimum encoded cover, it asserts the output combination given by the intersection of the codes of *all* the symbolic values. If this intersection is null (all 0's), then the cube can be discarded to obtain a smaller cover. If the intersection is not null and the cube asserts some outputs, then it means that for the bits corresponding to these outputs, all the codes of the symbolic values have a 1. One can reduce the codes of the all the values and still maintain their identities by discarding these outputs. Then, the cube asserts a null output combination and can be discarded. Thus, the cube is not required in a minimum cover.

Hence, a minimum cardinality encodeable selection can be made up entirely of GPIs. □

Thus, if one selects a minimum set of GPIs that cover all minterms and have an associated set of constraints by Eqn. 4.1 that is encodeable, one is guaranteed a minimum solution to the encoding problem. It remains to prove that the conditions to the satisfied by the graph for encodeability are necessary and sufficient conditions.

Theorem 4.2.2 *Conditions 1-4 stated in the section on dominance and disjunctive relationships are necessary and sufficient for the graph to be encodeable.*

For a proof of the above theorem, see [30].

4.2.6   Multiple Symbolic Outputs

The procedure outlined can be generalized to the case where there are multiple symbolic outputs. Each minterm initially has a number of tags equal to the number of symbolic outputs. Each tag corresponds to the symbolic value whose on-set the minterm belongs to, for each symbolic output. Minterm pairs are merged and the operations on the tags are performed exactly the same as before. A $k+1$-cube cancels a k-cube only if all of its tags are identical to the corresponding tags of the k-cube. Cubes with tags such that all corresponding symbolic values are contained in the tag can be discarded. Thus, the GPIs can be generated. Separate graphs are used for representing encoding constraints for each symbolic output. Given a selection of GPIs, these graphs can be constructed and checked for encodeability as before. All the graphs have to be encodeable for a selection of GPIs to be valid.

The generalization to functions with both symbolic and binary-valued outputs is described in Section 5.2.1 of the next chapter.

4.2.7   The Issue of the All Zeros Code

If a code of all zeros is given to a symbolic value, then it is possible that one or more GPIs can be dropped in a PLA implementation, from an otherwise minimum cover. This is because in a PLA implementation, one is only concerned with the on-sets. The procedure presented has not taken this fact into account.

A solution is to perform $n + 1$ minimizations where n is the number of symbolic values. One minimization is as before. In the other n minimizations, all the minterms in the on-set of each of the n symbolic values are dropped, one value's on-set at a time. The best solution out of the $n + 1$ minimizations is then selected. The reason one has to perform the first minimization without dropping any of the minterms is that the all zeros code cannot appear in disjunctive relations, since it is dominated by all other codes. Hence, constraining oneself to use a code of all zeros may result in a sub-optimal solution.

It can be proved that the following theorem gives conditions where multiple minimizations are not required.

Theorem 4.2.3 *Given a cover with one or more symbolic outputs and binary-valued outputs if all minterms in the cover belong to the on-set of at least one binary-valued output, then there can be no advantage to using an all zeros code.*

Proof. The only advantage in using an all zeros code is that minterms may be dropped by putting them into off-sets. One can always satisfy required dominance and/or disjunctive relationships via codes other than the all zeros code. In the case of a cover with the property mentioned above, one cannot drop any of the minterms. Hence, one can obtain a minimum cardinality solution without using the all zeros code. □

4.2.8 Reduced Prime Implicant Table Generation

Many techniques for determining all the PIs of logic functions with single and multiple outputs have been published in the past (*e.g.* [68] [90]). An algorithm based on the recursive decomposition of a function followed by a pair-wise consensus operation has been reported [13] and has been improved upon in the program MCBOOLE [21]. Other techniques have been reported in [80]. These techniques not only efficiently generate PIs without duplication of effort but also create a *reduced prime implicant table*. In the prime implicant table of the Quine-McCluskey (Q-M) algorithm, each column in the table corresponds to a minterm of the function and each row to a PI. In a reduced prime implicant table, each column corresponds to a collection of minterms (*i.e.* a larger subspace), all of which are covered by the same set of PIs. Thus, using the

0001	out1	0001	11110
00-0	out2	00-0	11101
0011	out2	0011	11101
0100	out3	0100	11011
1000	out3	1000	11011
1011	out4	1011	10111
1111	out5	1111	01111

(a)   **(b)**

Figure 4.14: Transformation for output encoding

algorithms of ESPRESSO-EXACT [80] for example, rather than the Q-M method leads to a more efficient creation of the prime implicant table.

These techniques cannot be used directly on functions with symbolic outputs to generate all GPIs. The canceling rule for GPIs is not the same as the canceling rule for PIs. However, one can transform a function with a symbolic output into a function with multiple binary-valued outputs such that the PIs for this new multiple-output function have a one-to-one correspondence with the GPIs of the original function. This is illustrated in Figure 4.14. The function with a symbolic output of Figure 4.1(b) has been duplicated in Figure 4.14(a). Each symbolic value is replaced by an output combination to produce the binary-valued multiple-output function of Figure 4.14(b). M outputs are required if there are M symbolic values. A symbolic value has an output combination of all 1's and one 0 in a unique identifying position. These outputs perform the same function as the output tag in GPI generation (*cf.* Section 4.2.1).

Lemma 4.2.2 *The PIs of the function obtained via the transformation described are the GPIs of the original function with the symbolic output.*

Proof. The set of outputs asserted by any cube in the new function is the set of symbolic values *not* in the tag of the corresponding cube in the original function. While generating the PIs for the binary-valued multiple-output function, a cube, c_1, cancels another cube, c_2, only if c_1 covers c_2 and the outputs asserted by c_1 are the same as the outputs asserted by c_2. This implies that the set of symbolic values in the tag of the two corresponding cubes in the original function are identical and c_1 would have canceled c_2 there as well. Finally, cubes in the binary-valued function formed with a null output combination are discarded. This corresponds to discarding cubes with tags containing all the symbolic values. □

Thus, via this transformation it is possible to make use of the classical techniques for prime implicant generation. Once the GPIs have been generated, the additional outputs are discarded, since a covering problem different from the standard covering problem has to be solved. The output tags for each GPI are constructed by finding all the symbolic values whose on-sets intersect the GPI.

4.2.9 Covering with Encodeability Constraints

The classical covering problem of two-level Boolean minimization involves finding a minimum set of prime implicants (PIs) that form a cover for a logic function. Here, there is the additional restriction on the selected generalized prime implicants (GPIs) − they have to form an *encodeable* cover. The definition of encodeability varies for the output encoding and state assignment. However, the covering algorithm need only be concerned with a black box that determines encodeability of the selected set of GPIs and a few other properties of the constraint graph associated with the selected GPIs.

The standard branch-and-bound solution to the minimum cover problem involves the following steps (rows correspond to PIs and columns to collections of minterms):

1. Remove columns that contain other columns and remove rows which are contained by other rows. Detect essential rows (a column with a single 1 identifies an essential row) and add these to the selected set. Repeat until no new essential elements are detected.

2. If the size of the selected set exceeds the best solution thus far, return from this level of recursion. If there are no elements left to be covered, declare the selected set as the best solution recorded thus far.

3. Heuristically select a branching row.

4. Add this row to the selected set and recur for the sub-table resulting from deleting the row and all columns that are covered by this row. Then, recur for the sub-table resulting from deleting this row without adding if to the selected set.

In ESPRESSO-EXACT, a lower bounding technique based on a maximal independent set heuristic was proposed. In Step 2, a maximal set of columns, all of which are pairwise disjoint is found using a straightforward, greedy algorithm (Finding a *maximum* independent set of columns is itself NP-complete [35]). Because each column must be covered and all the columns in the maximal independent set share no row in common, the size of the maximal independent set is a lower bound on the number of rows required to complete the cover. At Step 2, the recursion can be bounded if the size of the selected set at Step 2 *plus* the size of the maximal independent set equals or exceeds the best solution known.

The covering algorithm used in output encoding is a modification of the algorithm described above. The modifications are described in the sequel.

In Step 1, a row (GPI) is deemed to contain another row (GPI) only if the tags of the two GPIs are identical or the tag of the first GPI is a subset of the tag of the second. (This may happen lower in the recursion after some columns have been deleted.) The lower bounding criterion at Step 3 uses the size of the maximal independent set of columns. This bound is looser than in standard covering because even if a cover can be constructed with a number of elements equal to the lower bound, it may not be encodeable.

Once the selected set covers all elements, an encodeability check is performed. If the cover is encodeable, the best recorded until then is declared as the solution. If not, another branch-and-bound step is performed to find the *minimum* number of GPIs (rows) which when added to the selected set renders it encodeable. The GPIs during this branch-and-bound step are selected from the current sub-table in the recursion. This branch-and-bound step is now described.

1. If the selected set is encodeable, then declare the selected set as the best encodeable solution thus far. If not, check if the size of the selected set *plus* a lower bound on the required number of rows to

produce an encodeable set equals or exceeds the best encodeable solution obtained thus far. If so, return from this level of recursion.

2. Heuristically select a branching row.

3. Add this row to the selected set and recur for the sub-table resulting from deleting this row. Then, recur on deleting the row without adding it to the current set.

The concern in this branch-and-bound step is no longer to cover more minterms since all minterms have already been covered. The lower bound on the number of GPIs required to render the graph encodeable is estimated by finding the number of *disjoint* violations of the encodeability conditions of Theorem 4.2.2.

If there are two cycles in the graph such that the edges in cycle 1 have different labels from all the edges in cycle 2 and no unselected GPI exists that contains both minterms corresponding to the labels of any pair of edges, then two GPIs are required to break both cycles. These two cycles are disjoint cycles. Similarly, assume that there are two instances of directed paths between siblings of a disjunctive arc. If the two sets of edges in the two paths have disjoint sets of labels and no unselected GPI exists that covers the pair of minterms corresponding to any pair of edges in the two paths, then two GPIs are required to remove the two violations. Disjoint violations of Conditions 3 and 4 of Theorem 4.2.2 can exist as well.

The heuristic selection of a GPI to be added to the selected set at Step 2 is performed by selecting a GPI that covers a large number of minterms corresponding to the labels of edges that are involved in violations of the encodeability conditions.

4.2.10   Experimental Results Using the Exact Algorithm

In this section, experimental results given in [30] for exact solutions to the output encoding problem, that were obtained on a set of examples are presented.

The results obtained using the output encoding algorithm are given in Table 4.1. In the table, the number of inputs to the function (inp), the number of minterms in the original function (min), the number of symbolic values (val), the number of binary-valued outputs (out), the

EX	inp	min	val	out	gpi	prod	enc	CPU time
ex1	2	4	4	1	6	3	2	0.1m
ex2	4	15	6	1	23	6	3	0.9m
ex3	6	44	16	2	194	14	6	10.4m
ex4	8	113	20	0	950	50	9	53.6m
ex5	10	213	20	1	8807	-	-	> 1h
ex6	12	410	32	0	> 9999	-	-	-

Table 4.1: Results using the exact output encoding algorithm

number of GPIs generated (gpi), the number of product terms in the minimized result (prod), the number of encoding bits (enc) and the CPU time in minutes required for GPI generation, covering and encoding on a μVAX-III^{TM} (CPU time) are given for each example. For example *ex5*, the covering problem could not be solved in less than a CPU-hour. For example *ex6* all the GPIs could not be generated due to memory limitations. Examples *ex3* and *ex4* which have upto 20 symbolic values have been successfully encoded, for which an exhaustive search method is not feasible.

4.2.11   Computationally Efficient Heuristic Minimization

The exact algorithms described in the previous section may require inordinate amounts of memory or CPU time to run due to the following reasons:

1. The number of GPIs may be too large.

2. Checking for encodeability given a set of dominance and disjunctive relations can be accomplished in polynomial time by checking for the conditions of Theorem 4.2.2. However, there are a number of alternatives in choosing these relations. It is conjectured that the problem of checking if an arbitrary set of GPIs can satisfy Eqn. 4.1 via some encoding is NP-complete, since the number of equations can be exponential and each equation represents multiple choices.

3. The classical covering problem is NP-complete and there is an additional branch-and-bound step in the constrained covering problem.

The programs ESPRESSO-EXACT and MCBOOLE appear to be capable of exactly minimizing most encoded FSMs. However, functions like those of Figure 4.14(b) have huge numbers of PIs. This and having to continually check for encodeability are the reasons why the exact output encoding algorithm falls short of the performance of exact two-level logic minimization algorithms. Many heuristics based on the exact procedures described here have been proposed. These heuristics may result in sub-optimal solutions but are computationally efficient.

During GPI generation, one can discard (or not generate) GPIs with tags that contain more that $k \leq n - 1$ symbolic values, where n is the total number of symbolic values. If $k = 1$, a disjoint minimization problem has to be solved, equivalent to that of minimizing each of the symbolic value on-sets separately. The case where $k = n - 1$ corresponds to an exact output encoding problem. Reducing the number of symbolic values that can be contained in a GPI has two advantages. Firstly, the number of GPIs is reduced. Secondly, the encodeability check becomes easier because the constraints specified by Eqn. 4.1 are simpler.

Another strategy which can be used in conjunction with the above heuristic or in isolation is to define a *stronger* form of encodeability that is easier to check for. As long as the definition includes the conditions of Theorem 4.2.2, an encodeable solution will be obtained. The heuristic may miss an optimum solution because it may consider the solution not encodeable, when in reality it is encodeable. A stronger definition of encodeability would restrict the alternatives in satisfying a constraint given by Eqn. 4.1. Consider Eqn. 4.2. There are 7 possible choices in satisfying the constraint. If one restricted all relations to be dominance relations, there would be 3 choices in satisfying Eqn. 4.2. Similarly, if one restricted all relations to be disjunctive, there would be 4 alternatives. One can also spend a specified amount of time (or backtracks) checking for encodeability of a set of constraints and consider the selected set of GPIs to be not encodeable if a selection of edges representing a compatible set of relations has not been found within the prescribed limit.

```
0000  out1
0011  out1
0001  out2
0100  out2
0101  out3
1000  out4
1010  out1/out2/out3/out4
1011  out1/out2
```

Figure 4.15: Symbolic output don't cares

A third heuristic approach is based on the iterative optimization strategy to two-level logic minimization, first used in MINI [46]. One can begin with the given symbolic function represented as GPIs and iteratively reduce, reshape and expand the GPIs while maintaining coverage of the minterms *and* encodeability. Since the iterative optimization approach has met with great success in the heuristic two-level logic minimization area, it is felt that this approach holds the most promise. Such methods have been developed for the decomposition problem and will be described in Chapter 6.

4.3   Symbolic Output Don't Cares

Don't cares for binary-valued functions are simply represented and exploited in logic minimization. Functions with symbolic outputs may have associated don't care conditions with certain input combinations as well. We term these don't cares to be **symbolic output don't cares**.

A symbolic output don't care is defined on the set of symbolic values that the function can take. For instance, the minterm 1010 in the function of Figure 4.15 is a symbolic don't care. A symbolic output don't care may encompass all the symbolic values of the function or only a subset. Minterm 1011 of Figure 4.15 is a don't care which can take on only a subset of the complete set of symbolic values.

One can produce an exact solution to an output encoding problem (based on the method described in Section 4.2) under an arbitrary symbolic output don't care set as follows. Add the don't care minterms to the on-sets of each of the symbolic values that the minterm can take. GPIs are generated as before. However, we may have a situation where two identical k-cubes have tags such that the first one's tag is a subset of the other. In this case the first k-cube cancels the second.

Given all the GPIs, the covering problem is solved as before. The minterms corresponding to the symbolic output don't cares have to be covered as well and Eqn. 4.1 has to be satisfied for them, else they may assert an invalid binary combination in the encoded cover. However, Eqn. 4.1 for these minterms has more choices, since a minterm effectively belongs to multiple symbolic value on-sets (multiple v_m's in Eqn. 4.1). Any one of these constraints is to be satisfied. For example, we may have:

$$out1 \; \cap \; out2 \; \bigcup \; out1 \; \cap \; out3 \; = \; out1 \; \; or \; \; out2$$

for a symbolic output don't care.

There is one other subtle constraint associated with encodeability when symbolic output don't cares are specified. [2] If a certain GPI that contains a symbolic output don't care minterm is selected, other GPIs corresponding to a different tag being asserted by the same don't care minterm are not allowed in the selected cover.

Symbolic don't cares are easily incorporated into the Boolean satisfiability formulation of the encodeability problem which will be described in Section 5.2.7 of the next chapter. Lin and Somenzi in [62] give a detailed treatment of the formulation of the symbolic don't care minimization problem (called symbolic relations in [62]) as one of binate covering. The entire minimization problem *i.e.* generating generalized prime implicants, constrained covering with encodeability checking, as well as the issue of the all zeros code, is uniformly dealt with in [62].

4.4 Output Encoding for Multilevel Logic

There has not been much work done in output encoding specifically targeting targeting multilevel logic implementations. We will defer

---

[2]This constraint has some similarity to the constraints posed in the minimization of Boolean relations [16].

the description of state encoding strategies that model interactions in the next state field (the output plane), with a view to targeting multilevel logic, to Chapter 5.

4.5 Conclusion

Output encoding problems have proven to be more difficult to analyze, and solve optimally than their input encoding counterparts. Interactions in the output plane of a multiple-output Boolean function are more complex than those in the input plane. However, progress has been made recently in understanding the interactions in the output plane during the minimization of a multiple-output Boolean function. In particular, notions of dominance [71] and disjunctive [30] relations have allowed the development of efficient, heuristic strategies for two-level output encoding.

The notion of generalized prime implicants (GPIs) [30] has enabled the development of a strategy for symbolic output minimization. In conjunction with the definition of an encodeable cover, GPIs allow for a systematic means of solving the output encoding problem, exactly minimizing the number of product terms in the eventual implementation. However, substantial work needs to be done in order to raise the efficiency of symbolic output minimization to the efficiency-level of symbolic/multiple-valued input minimization.

Chapter 5

State Encoding

Work in the encoding of symbolic inputs and symbolic outputs was described in the two previous chapters. The state assignment problem is an input-output encoding problem with equality constraints on the symbolic inputs and outputs. In Figure 5.1, a State Transition Table (STT) of a finite state machine (FSM) is shown. The present states (2nd column) can be viewed as a symbolic input and the next states (3rd column) can be viewed as a symbolic output.

In this chapter, strategies developed for integrating input and output encoding methods are described. It is shown in this chapter how the techniques for the encoding of symbolic inputs and outputs can be unified into a technique for the optimal encoding of symbols that appear in the input as well as the output planes. We first consider targeting two-level logic implementations in Section 5.1. We describe several heuristic techniques for state assignment that are based on the input encoding

```
0  S1    S1  1
1  S1    S2  0
1  S2    S2  0
0  S2    S3  0
1  S3    S3  1
0  S3    S3  1
```

Figure 5.1: State Transition Table of Finite State Machine

```
0 100   100 1        0 100   100 1
1 100   010 0        1 100   010 0
1 010   010 0        1 010   010 0
0 010   001 0        0 010   001 0
1 001   001 1        - 001   001 1
0 001   001 1
```

(a)   **(b)**

Figure 5.2: Multiple-valued functions

strategies presented in Chapter 3. We introduce the notion of symbolic next state don't cares, which is similar to the notion of symbolic output don't cares (*cf.* Section 4.3), in Section 5.3. We consider encoding strategies that target multilevel logic implementations in Section 5.4.

5.1   Heuristic State Encoding Targeting Two-Level Logic

As described in Chapter 3, the input encoding problem in isolation can be solved by representing the symbolic input as a multiple-valued variable, where each distinct symbolic value represents a distinct value of the multiple-valued variable. Exact minimization of the resulting multiple-valued function produces a minimum cardinality multiple-valued cover. The merged input implicants in the minimized multiple-valued cover represent constraints that the binary codes assigned to the symbolic states have to satisfy, in order to produce an encoded binary cover with the same cardinality as the minimized multiple-valued cover. Any set of these input constraints can always be satisfied by some encoding (*cf.* Chapter 3).

To solve the state assignment problem exactly, one can treat the present state space as a multiple-valued variable and solve the resulting output encoding problem exactly. Modifications that are required to the strategy presented in Section 4.2 of Chapter 4 will be described in Section 5.2. We will consider heuristic strategies toward integrating input and output encoding in the sequel.

5.1.1 Approximating State Encoding as Input Encoding

One can approximate the state assignment problem as one of input encoding as in [73]. In this case, the given State Transition Table (STT) is encoded using a **one-hot code**. A one-hot code for a symbolic input was described in Section 3.2 of Chapter 3. Given a STT, both the present state and next state fields are given a one-hot code. As in the input encoding case, the unused codes can be used as don't cares in a binary-valued input, binary-valued output minimization, or a multiple-valued input, binary-valued output minimization can be carried out. The STT cover of Figure 5.1 has been represented as a multiple-valued function in Figure 5.2(a). The symbolic state $s1$ is the value 100, $s2$ is 010 and $s3$ is 001. Minimizing the function produces the result of Figure 5.2(b). The function has five product terms. Thus, ignoring the next state plane guarantees an encoded implementation that has no more than five product terms. However, if we code $s1$ with 01, $s2$ with 00 and $s3$ with 10, we will obtain a result with three product terms after minimization.

In general approximating the state assignment problem as one of input encoding may produce results far from optimum. A worst-case scenario for this approximation is encoding counter FSMs. Given a 2^p-state autonomous counter, multiple-valued input minimization results in a minimized cover with 2^p product terms. However, the optimal encoded implementation of a 2^p-state counter has $O(p^2)$ product terms!

5.1.2 Constructing Input and Dominance Relations

While generating input and dominance relations that have to be satisfied by the same encoding, the first concern is the encodeability of the combined set of relations. It was possible to state the unconditional existence of an encoding that satisfies an arbitrary set of input relations (*e.g.* one-hot code), and the unconditional existence of an encoding that satisfies an arbitary acyclic set of dominance relations (*cf.* Theorem 4.1.1). However, it is not possible to state the unconditional existence of an encoding that satisfies *both* a set of input relations and a set of acyclic dominance relations.

The reason becomes clear in the proof of the theorem below, which was given in [71].

Theorem 5.1.1 : *An arbitrary set of input relations and an acyclic set of dominance relations can be satisfied by the same encoding if and only if the following condition is satisfied: For any symbolic value tuple $s1$, $s2$ and $s3$ such that $s1 \succ s2$ and $s2 \succ s3$, no input relation should exist such that the position corresponding to $s1$ and $s3$ has a 1 and the position corresponding to $s2$ has a 0.*

Proof. Necessity: Assume a set of relations violates the condition and an encoding exists satisfying all relations. Recall from the definition of a super-cube that if the super-cube of the constraint has a 1 (0) in a particular position, it means all the symbolic values in the constraint have codes that have a 1 (0) in that position. The super-cube of the input constraint will either have a 1 in a bit where $s2$ has a 0, or will have a 0 in a bit where $s2$ has a 1. In the former case, $s2 \not\succ s3$ and in the latter $s1 \not\succ s2$ and therefore the output relations could not have been satisfied.

Sufficiency: Pick an arbitrary input constraint from the input constraint set. We group the states with 1's in the constraint into $Q1$ and the states with 0's into $Q0$. We are guaranteed via the condition above that if $q_1 \succ q_2$ and $q_2 \succ q_3$, then both q_1 and q_3 don't belong to $Q1$ or don't belong to $Q0$. If $q_1 \in Q1$ and $q_3 \in Q0$, then we append a column to the states' codes such that all $q \in Q1$ have a 1 and all $q \in Q0$ have a 0. If $q_1 \in Q0$ and $q_3 \in Q1$, then we append a column to the states' codes such that all $q \in Q0$ have a 1 and all $q \in Q1$ have a 0. This does not violate the dominance relations. After we satisfy all the input constraints, we can satisfy the dominance constraints by adding extra bits. The addition of these extra bits will not affect the satisfaction of the input constraints. □

We need to modify the symbolic output minimization procedure of CAPPUCCINO [71] (*cf.* Section 4.1.2) in order to apply it to the state encoding problem. The necessary modifications are stated below.

- The present state plane of the FSM to be encoded is replaced with a multiple-valued variable. We thus have a multiple-valued input function with a symbolic output and some binary-valued primary outputs (the primary outputs of the FSM).

- The input relations are derived by using multiple-valued input, binary-valued output minimization, effectively one-hot coding the next states.

- The on-sets and off-sets of each symbolic next state and each binary-valued primary output are computed. The on-sets and off-sets have multiple-valued input parts.

- The symbolic minimization loop of Figure 4.4 is used with the modification that the off-sets of the binary-valued outputs are always added to the OFF_i of a particular iteration. Further, given the set of input relations, we check to see if the condition stated in Theorem 5.1.1 is violated upon the addition of an edge between v_j and any symbolic value $v_k \notin J$. If so, we add the on-set of v_k to OFF_i.

5.1.3 Heuristics to Minimize the Number of Encoding Bits

The heuristic column-based encoding methods described in Section 3.3.2 lend themselves to easy generalization when both input and dominance relations are to be satisfied. In the sequel, we will assume the condition of Theorem 5.1.1 is met by the given set of input and dominance relations.

The heuristic used in CAPPUCCINO first compacts the set of input constraints using the techniques summarized in Section 3.3.2. An input constraint is heuristically selected, transposed to obtain a column vector, and added to the current encoding. Note that the column vector may have to be bitwise complemented in order to satisfy the output relations. All the input relations that are satisfied by the current encoding are then deleted.

The key in obtaining a minimal-length encoding is the selection of the input constraint. CAPPUCCINO computes a *satisfaction ratio* corresponding to each input constraint as to how many remaining input constraints will be satisfied if the transpose of this input constraint is added to the encoding. The constraint with the maximum satisfaction ratio is then selected. This heuristic appears to be fairly effective for small-moderate sized examples, though it is greedy, and has minimal look-ahead capabilities.

The use of simulated annealing has been advocated to solve the problem of satisfying input and dominance relations (as well as disjunctive relations) in [60]. Annealing-based constraint satisfaction has been shown to be effective, especially when there is a bound on the number of encoding bits.

5.1.4   Alternate Heuristic State Encoding Strategies

The generation of input and dominance relations proceeds in the state assignment program NOVA [94] proceeds with some variations in the heuristic selection of the symbolic next states in the procedure **select**() of Figure 4.4. A variation of the symbolic minimization loop of CAPPUCCINO as used in the state assignment program NOVA is described below.

The present state field and the next state field of the given FSM are both one-hot coded. The resulting encoded FSM is minimized. Note that there may be sharing between the binary-valued primary outputs and the symbolic next states. The ON_i for each symbolic next state corresponds to the disjointly minimized implicants, *leaving the binary-valued outputs unchanged*. The don't care set for the chosen symbolic next state v_i in a particular iteration is explicitly computed by using the ON_j of all the next states v_j to which there is no path from v_i in the dominance graph, G_D. The off-set is calculated by using the ON_j of all the next states v_j to which there is a path from v_i in the dominance graph, G_D. This strategy carries around the explicit description of the on-sets and off-sets of the binary-valued outputs, as they will appear in the final Boolean-minimized cover. This was the strategy used in the output encoding example of Figures 4.5 through 4.8.

NOVA restricts itself to using a minimum-length encoding during the constraint satisfaction step, and attempts to maximally satisfy the set of input and dominance constraints.

5.2   Exact State Encoding for Two-Level Logic

In this section, we will describe the extensions required to the output encoding strategy of Section 4.2 in Chapter 4 in order to arrive at a state assignment method that exactly minimizes the number of product terms in a two-level logic implementation. We describe the generation of generalized prime implicants (GPIs) in Section 5.2.1, the encodeability condition in Section 5.2.2 and the process of constructing an optimized cover in Section 5.2.3. In Section 5.2.4 we give a transformation similar to that described in Section 4.2.8 which allows the use of prime implicant generation algorithms developed for multiple-valued input, binary-valued output functions to generate GPIs for a State Transition Table description of a finite state machine. Necessary modifications to the covering step are described in Section 5.2.5. Heuristics to minimize the number of encoding bits given a compatible set of dominance, disjunctive and input relations are described in Section 5.2.6. A Boolean satisfiability formulation of constrained input-output encoding is provided in Section 5.2.7.

5.2.1 Generation of Generalized Prime Implicants

It is now required to encode a function with multiple binary-valued inputs, a single multiple-valued input, one symbolic output and multiple binary-valued outputs. Each minterm has a tag corresponding to the symbolic next state whose ON-set it belongs to. Each minterm also has a tag that corresponds to all the outputs asserted by the minterm.

Two minterms or 0-cubes can merge to form a 1-cube. Merging may occur between minterms with the same binary-valued part and different multiple-valued parts or uni-distant binary-valued parts and the same multiple-valued part. The next state tag of the 1-cube is the union of the next state tags of the two minterms. As in the Quine-McCluskey method, the binary-valued output tag of the 1-cube will contain only the outputs that both minterms asserted. A 1-cube can cancel a 0-cube if and only if their next state and binary-valued output tags are identical *and* their multiple-valued parts are identical. Thus, a 1-cube 1 011 (where the second term is a multiple-valued implicant) cannot cancel 1 001 even if their next state and output tags are identical. This is because the merging of the multiple-valued part represents an input constraint as described in Chapter 3. One exception is when the multiple-valued input part of the 1-cube contains all the symbolic states − in this case the implicant represents an input constraint that is satisfied by *any* encoding.

Generalizing to k-cubes, we have:

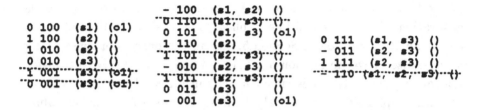

Figure 5.3: Generation of GPIs in state assignment

1. A $k + 1$-cube formed from two k-cubes has a next state tag that is the union of the two k-cubes' next state tags and an output tag that is the intersection of the outputs in the k-cubes' output tags.

2. A $k+1$-cube can cancel a k-cube only if their multiple-valued input parts are identical or if the multiple-valued input part of the $k + 1$-cube contains all the symbolic states. In addition, the next state and output tags have to be identical.

A cube with a next state tag containing all the symbolic states and with a null output tag can be discarded. The generation of GPIs for the FSM of Figure 5.1 is depicted in Figure 5.3. 13 GPIs are eventually produced.

5.2.2   Selecting a Minimum Encodeable Cover

Given all the GPIs, a minimum encodeable set is selected that covers all minterms by solving a covering problem (*cf.* Section 5.2.5), as before. However, the definition of encodeability is different due to the complication of having the input constraints.

An input constraint may conflict with dominance or disjunctive relations. Therefore, when a set of GPIs is picked, one needs to check that the input constraints, given by the merging of the multiple-valued implicants, as well as the relations given by Eqn. 4.1 are compatible. The question of compatibility between input constraints and dominance relationships was posed in Section 5.1.2 and a theorem stating necessary and sufficient conditions for compatibility was given. Here, a more complex case of possibly mutually conflicting input, dominance and disjunctive relations is encountered. For a proof of the theorem below see [30].

Theorem 5.2.1 *Given a set of dominance and disjunctive relations represented by a graph and a set of input relations, a necessary and sufficient set of conditions for the existence of an encoding satisfying all the relations are:*

1. *Conditions 1-4 of Theorem 4.2.2 are satisfied.*

2. *For any state tuple s1, s2 and s3 such that s1 ≻ s2 and s2 ≻ s3, no input relation should exist such that the position corresponding to s1 and s3 has a 1 and the position corresponding to s2 has a 0. This should hold even if s1 ≻ s2 is a disjunctive dominance relation, with s1 representing a disjunction of two or more states.*

3. *No input relation should exist where all the siblings (or ancestors of siblings) of a disjunctive equality arc have a 1 and the parent 0. In the case of the siblings being conjunctive nodes, no input relation should exist where all the immediate ancestors (or ancestors of the immediate ancestors) of each conjunctive sibling have a 1 and the parent 0.*

A selection of four GPIs for the example of Figure 5.1 is shown in Figure 5.4. The encodeability constraints for the six minterms in the STT of Figure 5.1 are also shown in Figure 5.4. Only two minterms have output constraints, the first of which is always satisfied. Therefore, we only have to satisfy the input relation 110 and the dominance relation $enc(s2) \succ enc(s3)$. The two relations are non-conflicting, as per the conditions of Theorem 5.2.1.

An elegant, dichotomy-based (*cf.* Section 3.3.4) algorithm that checks for the compatibility of an arbitrary set of input, dominance and disjunctive relations in polynomial time assuming an unbounded number of encoding bits was presented in [83]. Output constraints are incorporated into the dichotomy-based framework (*cf.* Section 3.3.4) as follows. Dominance constraints imply that if $enc(s1)$ dominates $enc(s2)$, all dichotomies with s1 in the left block and s2 in the right block are invalid (a symbol in the left block of a dichotomy is assigned a 0 in the bit associated with that dichotomy). Similarly, if $enc(s1)$ is the disjunction of $enc(s2)$ and $enc(s3)$, s1 should appear in the same block as either s2 or s3 in each dichotomy. Similar requirements are imposed

```
-  010   (S2,  S3)  ()
-  001   (S3)       (01)
0  100   (S1)       (01)
1  110   (S2)       ()
```

(a)

```
0  100
1  100
1  010    enc(S2) ∩ enc(S3) ∪ enc(S2) = enc(S2)
0  010    enc(S2) ∩ enc(S3) = enc(S3)
1  001
0  001
```

(b)

Figure 5.4: Encodeability constraints for a selected set of GPIs

when $enc(s1)$ is dominated by the disjunction of $enc(s2)$ and $enc(s3)$, and when nested conjunctions are involved in the disjunction.

Based on this approach, the algorithm for compatibility of the input and output constraints is as follows. From the initial set of dichotomies, the ones that violate an output constraint are removed. The output constraints also force symbols to be added to the remaining dichotomies. For example, if $enc(s1)$ dominates $enc(s2)$, then in each dichotomy that $s1$ appears in the left block, $s2$ is also inserted in the left block. This process of adding symbols to dichotomies based on the output constraints is called **raising** the dichotomies in [83]. When no further symbols can be added to a dichotomy, the dichotomy is said to be **maximally raised**. A dichotomy that is raised based on one output constraint could become incompatible with another output constraint. Such dichotomies are deleted. The various constraints are compatible if the set of maximally raised dichotomies covers the initial encoding dichotomies. The initial number of dichotomies is proportional to the number of input constraints. In order to maximally raise each initial dichotomy, each output constraint has to be processed once. Therefore the complexity of checking the compatibility of the constraints is of the order of the product of the number of input and output constraints.

Once the constraints are deemed compatible, prime dichotomies are generated from the set of maximally raised dichotomies. A minimum cardinality set of prime dichotomies that covers the initial di-

```
-  010  (S2, S3) ()                              -  11  10  0
-  001  (S3)     (01)       S1  ->  01          -  10  10  1
0  100  (S1)     (01)       S2  ->  11          0  01  01  1
1  110  (S2)     ()         S3  ->  10          1  -1  11  0
       (a)                      (b)                  (c)
```

Figure 5.5: Construction of optimized and state-assigned cover

chotomies corresponds to the minimum length encoding satisfying the input and output constraints. The minimum covering step is of course NP-complete.

The encodeability check for a set of GPIs, given a bound on the number of encoding bits that can be used, can also be formulated as a Boolean satisfiability problem. This formulation is given in Section 5.2.7.

5.2.3 Constructing an Optimized Cover

Once the GPIs have been selected and an encoding satisfying all relations is found, it a simple matter to construct the optimized cover. The output tag of each GPI gives the outputs asserted by the GPI. Intersecting the binary codes of all states in the next state tag gives the next state part (in binary form). The multiple-valued input part of a GPI is replaced by the super-cube of the codes of all states in the multiple-valued implicant.

In Figure 5.5, we have an encoding for the three states of the FSM of Figure 5.1 that satisfies the relations imposed by the selection of the four GPIs as was illustrated in Figure 5.4. We can therefore construct an encoded and minimized cover with four product terms (*cf.* Figure 5.5). Note that if we had used the all zeros code (*i.e.* 00) for $s2$, we could have obtained a three product term solution.

Arguments very similar to those of Section 4.2.5 can be used to show that the procedure described in this section does indeed result in a minimum solution to the state assignment problem.

5.2.4 Reduced Prime Implicant Table Generation

It is possible to transform the given State Transition Table (STT) into a multiple-valued input, binary-valued output function such

```
0 S1   S1 1              0 001    1 110 110
1 S1   S2 0              1 001    0 101 110
1 S2   S2 0              1 010    0 101 101
0 S2   S3 0              0 010    0 011 101
1 S3   S3 1              1 100    1 011 011
0 S3   S3 1              0 100    1 011 011
```

(a)   (b)

Figure 5.6: Transformation for state assignment

that the generalized prime implicants for the STT correspond to the prime implicants (PIs) of the binary-valued output function.

In the state assignment case, given the multiple-valued input variable, we have a restriction during GPI generation that the multiple-valued part of a $k+1$-cube that cancels a k-cube has to be identical. This does not apply to PI generation in multiple-valued input, binary-valued output functions [81]. Therefore, a more complex transformation is required in the case of a function with a symbolic input and output, as opposed to the transformation illustrated in Figure 4.14 of Chapter 4. This transformation is illustrated in Figure 5.6. In Figure 5.6(a), the STT of Figure 5.1 has been duplicated. The new function of Figure 5.6(b) has three sets of binary-valued outputs. The first set corresponds to the original binary-valued outputs in the STT. The second set corresponds to the next states. Given N_S states, there are N_S binary-valued outputs in this set. This set performs the function of the next state tag in GPI generation (*cf.* Section 5.2.1). The third and last set of outputs incorporates the restriction of the equality of the multiple-valued input parts for cube cancellation. This set of N_S outputs corresponds to the present state space. It is constructed like the second set − each state has a unique N_S-bit code with $N_S - 1$ 1's and one 0.

The argument that generating the PIs for this transformed function is equivalent to generating the GPIs for the original function follows in a similar way to the proof of Lemma 4.2.2.

Once the GPIs have been generated, the additional outputs are discarded, since a covering problem different from the standard covering problem has to be solved. The next state tags for each GPI are constructed by finding all the symbolic values whose ON-sets intersect the GPI.

5.2.5 The Covering Step

The covering step in output encoding was described in Section 4.2.9 of Chapter 4. The covering step for the state assignment problem is similar, and we will merely describe the modifications necessary to the strategies of Section 4.2.9 in the sequel.

In addition to the pruning techniques described in Section 4.2.9, we can tighten the lower bound on the number of GPIs required to render the graph encodeable by finding the number of disjoint violations of the encodeability conditions of Theorem 5.2.1. Disjoint violations of Condition 2 of Theorem 5.2.1 would have 2 state-tuples with dominance edge pairs that have different pairs of labels, with the restriction that no unselected GPI exists that covers the pair of minterms corresponding to any pair of edges. Similarly, one can have disjoint violations of Condition 3 of Theorem 5.2.1.

5.2.6 Heuristics to Minimize the Number of Encoding Bits

Heuristics were described in Section 3.3.2 to satisfy a set of input relations with a minimal number of bits. The extensions of column-based input encoding methods to include dominance relations were described in Section 5.1.2. The following procedure [30] is based on the procedures described in Section 3.3.2.

1. Compact the set of input relations using the fact that some input relations may be implied by others. In particular, an input relation that is the bitwise AND of two other input relations, is guaranteed to be satisfied if the two other relations are satisfied. An input relation that is the bitwise complement of a second input relation is guaranteed to be satisfied if the second input relation is satisfied (*cf.* Section 3.3.2).

2. Represent the reduced set of input relations by a matrix, where each column corresponds to a symbolic value and each row to a

constraint. Construct an encoding as the transpose of the matrix, *i.e.* each symbolic value/node receives as a code the column corresponding to the node in the original matrix. This encoding is guaranteed to satisfy the input relations (*cf.* Section 3.3.2).

3. Find the set of dominance relations between each pair of nodes that are not satisfied. No dominance relation could have been violated. Select a maximal disjoint set of pairwise dominance relations (By disjoint, it is meant that the two nodes in the dominance relation are distinct from the nodes in the other dominance relation). Satisfy these relations by adding a single bit to the encoding. Do so till all dominance relations are satisfied.

4. Disjunctive relations have to satisfied for each bit in the encoding. If for a given disjunctive equality arc, a bit in the codes corresponding to the parent/siblings in the arc violates the relation, it can only be because the bitwise OR of the siblings is a 0 and the parent is a 1 (This is because all the dominance relations have been satisfied). The choices of raising the bits to a 1 (from a 0) are tried in each possible subset of the siblings. At least one of these choices will not violate the dominance relations. However, an input relation may be violated and/or a dominance relation may no longer be satisfied.

5. For the input relations that are not satisfied, append a set of bits corresponding to the transpose of the compacted set of relations. Go to Step 3.

The procedure will converge since the set of relations is compatible.

5.2.7 Encoding Via Boolean Satisfiability

The problem of determining encodeability for a selection of GPIs and finding an encoding *within a certain length* that satisfies the constraints specified by Eqn. 4.1 can be formulated as a Boolean satisfiability problem. It should be noted that Theorem 4.2.2 of Chapter 4 and Theorem 5.2.1 give conditions for a graph obtained via *a particular selection of edges* to be encodeable. Hence, to determine encodeability of a set of constraints given by Eqn. 4.1, one has to effectively try all possible selections.

Satisfying Eqn. 4.1 can be viewed as satisfying a Boolean expression. Given a cover with a set of symbolic values v_0, v_1, .. v_{n-1} and a bound on the number of encoding bits that can be used, B, one can construct a logic function corresponding to the encodeability of a selection of GPIs. If the logic function is satisfiable, then the selection of GPIs is encodeable *and* an encoding can be determined from any minterm that satisfies the logic function.

Each of the v_i is represented a set of B distinct variables l_{ij}, $1 \leq j \leq B$. It is necessary to satisfy the constraint that the vectors corresponding to the l_{ij} have to be different for all i. This is accomodated by writing the Boolean expressions:

$$l_{i1} \oplus l_{k1} + l_{i2} \oplus l_{k2} + .. + l_{iB} \oplus l_{kB} \quad 0 \leq i, k < n, i \neq k \quad (5.1)$$

Each of these expressions has to evaluate to a 1 ($\oplus$ is the exclusive-or operation). The Boolean expressions corresponding to Eqn. 4.1 are the equations themselves with $\cap$ replaced by a bitwise AND, $\cup$ replaced by an OR and = replaced by an exclusive-nor. For example, an equation $enc(v_1) \cap enc(v_2) \cup enc(v_1) \cap enc(v_3) = enc(v_1)$ becomes a Boolean expression:

$$((l_{11}l_{21} + l_{11}l_{31}) \otimes l_{11}) .. ((l_{1B}l_{2B} + l_{1B}l_{3B}) \otimes l_{1B}) \quad (5.2)$$

where $\otimes$ is the exclusive-nor operation. These Boolean expressions also have to evaluate to a 1 to satisfy Eqn. 4.1. Thus, it is required to find 0/1 values for all the l_{ij} such that Eqns. 5.1 and all Eqns. 5.2 evaluate to a 1. If such a set of values can be found, then we have an encodeable set of GPIs and an encoding for the symbolic values.

The state assignment case has a more complex formulation, since we have to deal with input constraints as well as output relations. It is possible to write a Boolean equation to check if an input constraint is satisfied by a given encoding. Assume that there are n symbolic states v_0, .. v_{n-1}. Each v_i is represented by a set of B distinct variables l_{i1}, l_{i2}, .. l_{iB} as before. Given an arbitrary input constraint, let the states that are in the constraint be v_{t_1}, .. v_{t_T} and the states that are not in the constraint be v_{r_1}, .. v_{r_R}. The input constraint can be written as:

$$(l_{t_11} \oplus l_{r_k1})(l_{t_21} \oplus l_{r_k1}) .. (l_{t_T1} \oplus l_{r_k1}) \ +$$
$$(l_{t_12} \oplus l_{r_k2})(l_{t_22} \oplus l_{r_k2}) .. (l_{t_T2} \oplus l_{r_k2}) \ +$$

$$.. (l_{t_1 B} \oplus l_{r_k B})(l_{t_2 B} \oplus l_{r_k B}) .. (l_{t_T B} \oplus l_{r_k B}) \quad 1 \le k \le R \quad (5.3)$$

Each of the above R equations has to evaluate to a 1. Such equations can be written for all the non-trivial constraints in a selected GPI set. Eqns. 5.1, Eqns. 5.2 and Eqns. 5.3 all have to be satisfied in order for a set of GPIs to be encodeable.

5.3   Symbolic Next State Don't Cares

Symbolic next state don't cares can be specified for the symbolic output of the given STT, similar to the symbolic output don't cares described in Section 4.3 of Chapter 4. Symbolic next state don't cares represent a method to take into account equivalences between symbolic states in a FSM that are to be encoded. Given an initial STT, that contains equivalent states, one can merge two (or more) equivalent states by giving them the same code, or give them distinct codes, corresponding to a state split machine. It has been shown [43] that using state splitting can reduce the area of a logic-level implementation.

Equivalent states correspond to a symbolic don't care condition – the next states of the fanin edges to state s_1 that is equivalent to state s_2 can be specified as (s_1, s_2). In the encoded and optimized machine, all or a subset of fanin edges to s_1 may be moved to s_2. If a given selection of GPIs results in all of the fanin edges of s_1 being moved to s_2, then we need not select GPIs to cover the fanout edges of s_1. This simply corresponds to utilizing the degree of freedom that s_1 and s_2 can be given the same code, which requires a small modification to the GPI generation and encodeability checking procedures described.

5.4   State Encoding for Multilevel Logic

5.4.1   Introduction

With the increasing use of multilevel logic in VLSI design, and multilevel logic optimization in VLSI synthesis, it became apparent that the cost functions corresponding to targeting two-level logic implementations of encoded finite state machines (FSMs) had to be extensively modified in order to obtain optimal multilevel logic implementations.

The first documented effort in FSM encoding that directly targeted multilevel logic implementations appears to be that of Devadas *et al* in [26]. An example of an optimal encoding for a two-level implementation resulting in a suboptimal multilevel implementation is presented in [26]. An approach to encoding with some similarities to the early work by Armstrong [1] is taken. Static estimations of the *encoding affinity* between states are made, based on literal savings estimates. That is, the literal savings that can potentially occur during multilevel logic optimization, if a given pair of states are coded with uni-distant codes is heuristically estimated. A weighted graph is constructed whose nodes are the states in the FSM, and whose edges carry weights corresponding to the estimated literal savings in coding the pair of states connected by any given edge with uni-distant codes. The actual encoding is done in a graph embedding step, where the states are assigned codes that minimizes the sum of the distance between each pair of states times the weight of the edge between them. This approach was fairly successful in modeling the common cube extraction step in multilevel logic optimization, but it was not very effective in modeling common subexpression extraction. The algorithms of [26] were implemented in the program MUSTANG and better results were obtained, in general, over the two-level encoding program KISS [73]. Some basic processes relating the coding of states to common cubes in the encoded implementation are described in Section 5.4.2. The two literal savings estimation algorithms of [26] are described in Section 5.4.3, and Section 5.4.4, respectively. The graph embedding step is described in Section 5.4.5.

Improvements to the MUSTANG literal savings estimation strategy in the next state and output plane, and the embedding strategy were made in the programs JEDI [61] and MUSE [33]. The description of these improvements is given in Section 5.4.6. Work done in input encoding targeting multilevel logic has already been described in Section 3.4.

In the sequel, the number of primary inputs to the machine is denoted N_i, the number of states N_s, the number of encoding bits N_b, and the number of outputs is denoted N_o.

5.4.2 Modeling Common Cube Extraction

There are two basic processes behind the influence of state assignment on the number of common cubes in the encoded State Transi-

–0	st0	st0	0
11	st0	st0	0
01	st0	st1	–
0–	st1	st1	1
11	st1	st0	0
10	st1	st2	1
1–	st2	st2	1
00	st2	st1	1
01	st2	st3	1
0–	st3	st3	1
11	st3	st2	1

Figure 5.7: Example Finite State Machine to be state assigned

tion Table (STT), a two-level representation, which is the starting point for multi-level logic optimization.

To begin, consider the second field (the present state field) in the STT of the machine shown in Figure 5.7. If the states $st0$ and $st2$ are assigned codes of distance N_d, then the lines of the next state $st1$ will have a common cube with $N_b - N_d$ literals (due to edges 3 and 8 in the STT). Similar relationships exist between other sets of states.

Now consider the third field (the next state field) of the STT. Let us assign the states $st0$ and $st2$ with codes of distance distance N_d. In this case, the present state $st1$ becomes a common cube for $N_b - N_d$ next state lines *whatever its code is* (due to edges 5 and 6 in the STT). The number of literals in the common cube is, of course, N_b. Again, similar relationships exist between other sets of states in the machine.

The input and output spaces (the first and fourth fields) also have an influence on the number of common cubes after encoding. If two different input combinations, i_1 and i_2, produce the same next state from different or same present states, then there is a common cube corresponding to $i_1 \cap i_2$ in the input space. Similarly, outputs asserted by different present states have common cubes corresponding to their intersections.

Given any machine, there are a large set of relationships between state encoding and the number/size of common cubes in the network prior to logic optimization. The reduction in literal count or "gains" that can be obtained by coding a given pair of states with close codes, so single/multiple occurrences of common cubes can be extracted, have to be estimated. Given these gains for each pair of states, one can attempt to find an encoding which maximizes the overall gain.

There arise complications in gain estimation. First, while the number of literals in the common cubes can be found exactly, the number of *occurrences* of these cubes in the logic function depends on the encoding of the next states. In the example above, assume that $st0$ was assigned 111 and $st1$ was assigned 110. There is a common cube 11 (with 2 literals) for the next state lines but the number of occurrences of this common cube depends on the number of 1's in the code of $st4$ (which is not known at this time). This problem is alleviated by treating the gains as *relative* merits rather than absolute and using an average-case analysis (see Section 5.4.3).

The second complication is that these statically-computed gains interact. Extracting some common cubes can increase the number of logic levels (to the outputs) of other common cubes and can also decrease the gain in extracting them. For instance, a sequence of two cube extractions on a two-level network can produce a three or a four-level network. Statically computing gains and maximizing the number of common cubes is fairly effective because, given a particular encoding, the optimal sequence of cube extractions to produce a minimal-literal multi-level network can be found by the logic optimizer.

The approach used in MUSTANG is to build a graph $G_M(V, E_M, W(E_M))$ where V, the set of nodes in G_M, has a one-to-one correspondence with the states of the finite state machine, E_M is a complete set of edges, *i.e.* every node is connected to every other node, and $W(E_M)$ represents the gains that can be achieved by coding the states joined by the corresponding arc as close as possible. These gains are statically and independently computed by enumerating the different relationships between the input, state and output spaces.

Then, the states are encoded, using this graph to provide the cost of an assignment of a state to a vertex of the Boolean hypercube (*cf.* Section 5.4.5).

A critical part of this approach is the generation of $W(E_M)$. Two algorithms will be described here: one assigns the weights to the edges by taking into consideration the second and fourth fields of the State Transition Table, and is henceforth called *fanout-oriented*. The

second algorithm assigns weights to the edges by taking into consideration the first and third fields and is henceforth called *fanin-oriented.*

The fanout-oriented algorithm attempts to *maximize the size*
of the most frequently occurring common cubes in the encoded machine prior to optimization. The fanin-oriented algorithm attempts to
maximize the number of occurrences of the largest common cubes in
the encoded machine prior to optimization. These two algorithms are
based on the two different processes behind the influence of state assignment on the number of common cubes in the network described
earlier. The algorithms define a set of weights for the undirected graph
$G_M(V, E_M, W(E_M))$. The weights represent a set of closeness criteria
for the states in the machine which reflect on the number of common
cubes in the encoded machine prior to optimization. Both these algorithms have a time and space complexity polynomial in the number of
inputs, outputs and states in the machine to be encoded. In the next
two sections, the two algorithms are described and analyzed.

5.4.3   A Fanout-Oriented Algorithm

This algorithm works on the output and the fanout of each
state. Present states which assert similar outputs and produce similar
sets of next states are given high edge weights (and eventually close
codes) so as to maximize the size of common cubes in the output and
next state lines.

The algorithm proceeds as follows:

1. Construct a complete graph $G_M(V, E_M, W(E_M))$, with the edge
   weight set, $W(E_M)$ empty. For each output, all the edges, E, in
   the State Transition Graph G, are scanned to identify the nodes
   which assert that output. N_o sets of weighted nodes which assert
   each output are constructed. If a node asserts the same output
   more than once it has a correspondingly larger weight in the set.

2. For each next state, sets of present states producing that next state
   are found. N_s sets are constructed. Steps 1 and 2 of the fanout-
   oriented algorithm are described in the pseudo-code of Figure 5.8.

3. Using these N_o $OSET$ and N_s $NSSET$ sets of nodes with their
   appropriate weights, $W(E_M)$ is constructed. The edge weight,

```
for ( i = 1; i ≤ N_o; i = i + 1) {
    foreach ( edge e ∈ E ) {
        if ( e.Output[i] is 1 ) {
            OSET_i = OSET_i ∪ e.Present ;
            Increment ow(i, e.Present) by 1 ;
        }
    }
}
foreach ( edge e ∈ E ) {
    NSSET_{e.Next} = NSSET_{e.Next} ∪ e.Present ;
    Increment nw(e.Next, e.Present) by 1 ;
}
```

Figure 5.8: Steps 1 and 2 of the fanout-oriented algorithm

```
foreach ( e_M(v_k, v_l) ∈ E_M ) {
    foreach ( v_i ∈ V )
        Increment we(e_M(v_k, v_l)) by nw(v_i, v_k) × nw(v_i, v_l) ;
    Multiply we(e_M(v_k, v_l)) by N_b/2 ;
    for ( i = 1; i ≤ N_o; i = i + 1 )
        Increment we(e_M(v_k, v_l)) by ow(i, v_k) × ow(i, v_l) ;
}
```

Figure 5.9: Step 3 of the fanout-oriented algorithm

we, is equal to the multiplication of the weights of the two nodes corresponding to the edge across all the sets. The weights corresponding to the next state sets have a multiplicative factor equal to the half the number of encoding bits, $\frac{N_b}{2}$. The reasoning behind the use of a multiplicative factor is given at the end of the section. The pseudo-code for the calculation of we is shown in Figure 5.9.

Analysis of the algorithm: The first step of the fanout-oriented algorithm entails enumerating the relationships between the present states and the output space. If two different present states assert an output, it is possible to extract a common cube corresponding to the intersection of the two state codes. By constructing the N_o different output sets and counting the number of times a pair of states occurs together in each output set, the algorithm effectively computes the number of occurrences of the common cube $X \cap Y$, for all states X and Y.

In the second step, the next states produced by each pair of present states are compared. A state pair which produces the same next state has an associated common cube corresponding to the pairwise intersection. The number of occurrences of this common cube is dependent on the number of 1's in the code of the next state and therefore cannot be estimated exactly (unlike in the first step). It is assumed that the average number of 1's in a state's code is $\frac{N_b}{2}$. Since one is concerned with relative rather than absolute merits, the approximation that each common cube occurs in $\frac{N_b}{2}$ next state lines is a fairly good one. Thus, a multiplying factor of $\frac{N_b}{2}$ is used in the second step. Ideally, this factor should be a function of the encoding and not a constant for all state pairs.

Given the number of occurrences of different common cubes in the machine, this algorithm assigns weights so as to maximize the size of the most frequently occurring cubes.

An example: The graph generated by the fanout-oriented algorithm for the example FSM of Figure 5.7 is shown in Figure 5.10. The output set corresponding to the single output is $(st0^2, st1^3, st3^2)$, where the superscripts denote the weights $nw()$ for each state in the set. The next state sets are:

$$st0 \rightarrow (st0^2, st1^1)$$

$$st1 \rightarrow (st0^1, st1^1, st2^1)$$

$$st2 \rightarrow (st1^1, st2^1, st3^1)$$

$$st3 \rightarrow (st2^1, st3^1)$$

The weight of the edge between the states $st2$ and $st3$ with $N_b = 2$ is $(1 \times 1 + 1 \times 1) \times \frac{N_b}{2} + 3 \times 2 = 8$. Similarly, the other edge weights can be calculated.

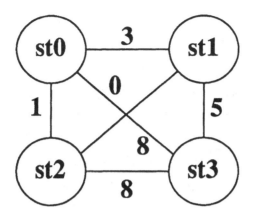

Figure 5.10: Graph produced by fanout-oriented algorithm

foreach (edge $e \in E$) {
    $PSSET_{e.Present} = PSSET_{e.Present} \cup e.Next$;
    Increment $pw(e.Present, e.Next)$ by 1 ;
}

Figure 5.11: Step 1 of the fanin-oriented algorithm

5.4.4 A Fanin-Oriented Algorithm

The fanout-oriented algorithm described above ignored the input space of the finite state machine. The fanout-oriented algorithm works well for FSMs with a large number of outputs and a small number of inputs. However, the number of input and output variables could both be quite large. In this section, a fanin-oriented algorithm is described, which operates on the input and fanin for each state. Next states which are produced by similar inputs and similar sets of present states are given high edge weights (and eventually close codes) so as to maximize the number of common cubes in the next state lines.

The algorithm proceeds as follows:

1. The graph G_M is constructed. N_s sets of weighted next states which fan out from each present state in G are constructed as shown in Figure 5.11.

```
for ( i = 1; i ≤ Ni; i = i + 1 ) {
    foreach ( edge e ∈ E ) {
        if ( e.Input[i] is 1 ) {
            ISET^ON_i = ISET^ON_i ∪ e.Next ;
            Increment iw(i, ON, e.Next) by 1 ;
        }
        if ( e.Input[i] is 0 ) {
            ISET^OFF_i = ISET^OFF_i ∪ e.Next ;
            Increment iw(i, OFF, e.Next) by 1 ;
        }
    }
}
```

Figure 5.12: Step 2 of the fanin-oriented algorithm

2. For each input, sets of next states are identified which are produced when the input is 1 and when the input is 0. $2 \times N_i$ such sets are constructed as shown in Figure 5.12.

3. The weights on the edges in the graph, we, are found using the N_i $ISET^{ON}$, N_i $ISET^{OFF}$ and N_s $PSSET$ sets of nodes as illustrated in the pseudo-code of Figure 5.13. Between each pair of nodes in G_M, an edge with weight equal to the multiplication of the weights of the two nodes across all the present state sets (scaled by N_b) and all the input sets is added.

Analysis of the algorithm: The first step of the fanin-oriented algorithm entails enumerating the relationships between the input and next state space. A next state produced by two different input combinations i_1 and i_2 has a common cube $i_1 \cap i_2$. The size of this cube can be found. By constructing the $2 \times N_i$ different input sets and counting the number of times a pair of states occurs together in each input set, the algorithm computes similarity relationships between all next state pairs in terms of the inputs. Giving next state pairs that are produced by similar inputs high edge weights will result in maximizing the number of occurrences of the largest common input cubes in the next state lines.

foreach ($(v_k, v_l) \in G_M$) {
    foreach ($v_i \in V$)
        Increment $we(e_M(v_k, v_l))$ by $pw(v_i, v_k) \times pw(v_i, v_l)$;
    Multiply $we(e_M(v_k, v_l))$ by N_b ;
    for ($i = 1;\ i \leq N_i;\ i = i + 1$) {
        Increment $we(e_M(v_k, v_l))$ by
            $iw(i, ON, v_k) \times iw(i, ON, v_l)$ +
            $iw(i, OFF, v_k) \times iw(i, OFF, v_l)$;
    }
}

Figure 5.13: Step 3 of the fanin-oriented algorithm

In the second step, the present states producing each pair of next states are compared. If two different next states are produced by the same present state, the state is common to some next state lines. The number of occurrences of this common cube is dependent on the intersection of the two next state codes. To maximize the number of occurrences of these cubes, next state pairs which have many common present states are given correspondingly high edge weights. Since each of these cubes have N_b literals (as opposed to a single literal for a single input), a multiplying factor of N_b is used while combining the weights computed in the two steps.

Given the sizes of the different common cubes in the machine, this algorithm assigns weights so as to maximize the number of occurrences of these cubes.

An example: The graph generated by the fanin-oriented algorithm for the example FSM of Figure 5.7 is shown in Figure 5.14. As can be seen, the weights of the edges in the graph are different from those generated by the fanout-oriented algorithm (Figure 5.10). Here the input sets are:

$$i_1(0) \rightarrow (st1^3, st3^2)$$

$$i_1(1) \rightarrow (st0^2, st2^3)$$

$$i_2(0) \rightarrow (st0^1, st1^1, st2^1)$$

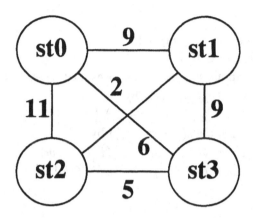

Figure 5.14: Graph produced by fanin-oriented algorithm

$$i_2(1) \rightarrow (st0^2, st1^1, st2^1, st3^1)$$

The present state sets are:

$$st0 \rightarrow (st0^2, st1^1)$$

$$st1 \rightarrow (st0^1, st1^1, st2^1)$$

$$st2 \rightarrow (st1^1, st2^1, st3^1)$$

$$st3 \rightarrow (st2^1)$$

The weight of the edge between $st0$ and $st1$ for $N_b = 2$ is $(1 \times 1 + 2 \times 1) + N_b \times (2 \times 1 + 1 \times 1) = 9$. The other edge weights are calculated in a similar fashion.

5.4.5 The Embedding Algorithm

The algorithms presented above generate a graph and a set of weights, like the graphs of Figure 5.10 and Figure 5.14, to guide the state encoding process. The problem now is to assign the actual codes to states according to the analysis performed by the fanin and the fanout-oriented algorithms. This problem is a classical combinatorial optimization problem called *graph embedding* [35]. Here, G_M has to be embedded in a Boolean hypercube so that the adjacency relations identified by G_M are satisfied in an optimal way. Unfortunately, this problem is NP-complete and there is little hope to solve it exactly in an efficient way. Thus, a heuristic approach to this embedding problem has to be used.

The heuristic algorithm described in [26] is called *wedge clustering*. This algorithm is used to assign codes to the nodes in G_M to minimize:

$$\sum_{i=1}^{N_s} \sum_{j=i+1}^{N_s} we(e_M(v_i,\ v_j)) \times dist(\ enc(v_i),\ enc(v_j)\ )$$

where the v_k are the vertices in G_M, $we(e_M(v_k,\ v_l))$ is the weight of the edge, e, between vertices v_k and v_l, and $enc(v_k)$ is the encoding of vertex v_k. The function $dist()$ returns the distance between two binary codes.

The graphs generated by the fanout and fanin-oriented algorithms have a certain structure associated with them, especially for large machines. In these graphs, typically small groups of states exist that are strongly connected internally (edges between states in the same group have high weights) but weakly connected externally (edges between states not in the same cluster have low weights). The embedding heuristic has been tailored to meet the requirements of this particular problem. The heuristic exploits the nature of the graph by attempting to identify strongly connected clusters and assigning states within each cluster with uni-distant codes.

The embedding algorithm proceeds as follows. Clusters of nodes with the cardinality of the cluster no greater than $N_b + 1$ and consisting of edges of maximum total weight are identified in G_M. Given G_M, the identification of these clusters is as follows − A node, $v_1 \in G_M$, with the maximum sum of weights of any N_b connected edges is identified. The N_b nodes, y_1, y_2, .. y_{N_b} which correspond to the N_b edges from v_1 and v_1 itself are assigned minimally distant codes from the pool of unassigned codes (v_1 may have been assigned already, so may the other y_i). A maximum of N_b nodes are chosen so the y_i can be (possibly) assigned uni-distant codes from v_1. After the assignment, v_1 and all the edges connected to v_1 have been deleted from G_M and the node selection/code assignment process is repeated till all the nodes are assigned codes. The pseudo-code of Figure 5.15 illustrates this procedure.

A local-optimality result for certain types of graphs can be proven for the above algorithm [26].

$GG = G_M$;
while (GG is not empty) {
    Select $v_1 \in GG$, $y_i \in GG$ such that
        $\sum_{i=1}^{N_b} we(e_M(v_1, y_i))$ is maximum ;
    Assign the y_i and v_1 minimally distant codes ;
    $GG = GG - v_1$;
}

Figure 5.15: Procedure for heuristic graph embedding

5.4.6 Improvements to Estimation Strategies

The experimental results obtained by the program MUSTANG validated the targeting of a different cost function, and the methodology used, since better results were obtained compared to the state-of-the-art two-level encoding programs at the time. However, the relatively simplistic modeling of the logic transformations in multilevel minimization left much to be desired.

Attempts were made to model common kernel extraction during state encoding (*e.g.* [95, 96]). Unfortunately, these initial attempts produced results worse than MUSTANG.

The calculation of graph weights in the program JEDI [61] is different from MUSTANG. In particular, rather than the $\times$ operator used to multiply the $nw()$, $pw()$, $ow()$ or $iw()$ weights in Step 3 of the fanout and fanin-oriented methods, the $+$ operator is used. Further, for state assignment, the next state field of the FSM is temporarily one-hot coded (*cf.* Figure 5.2) and a weighted graph $G^P{}_M$ for the present states is constructed using the fanout-oriented algorithm with the above variation. Next, the present states of the FSM are temporarily one-hot coded, and a weighted graph $G^N{}_M$ for the next states is constructed using a variation of the fanin-oriented algorithm. The two graphs are then combined into a single graph G_M whose edge weights correspond to the sum of the corresponding edge weights in $G^P{}_M$ and $G^N{}_M$.

The embedding algorithm in MUSTANG was improved upon in JEDI, where simulated annealing for used to solve the final graph embedding problem. On an average 20% better results were obtained by

JEDI over MUSTANG over a set of benchmark examples.

Du *et al* in [33] revisited the MUSTANG and JEDI strategies for encoding, and improved the literal-savings estimation part of the procedure. The estimations are based on analyzing algebraically-factored, one-hot coded implementations. Slightly better results over JEDI were obtained by MUSE.

Neither JEDI nor MUSE provide much additional insight into the modeling of multilevel logic transformations, other than the operation of common cube extraction. However, they do produce better multilevel logic implementations than other state assignment programs that are currently available.

5.5   Conclusion

State assignment is one of the oldest problems in automata theory. A large body of work exists in heuristic state assignment of finite state machines dating back to the '50s. With the advent of efficient two-level and multilevel logic optimization in recent years, the range of cost functions targeted by state encoding programs has become wider.

The prediction of the following logic minimization step is the crux of the optimal encoding problem. Substantial progress has been made in modeling the two-level logic minimization step at the symbolic level using multiple-valued input minimization for the symbolic inputs, and the notion of generalized prime implicants for the symbolic outputs. An exact state encoding method targeting the minimum number of product terms in an eventual two-level implementation has been proposed. Heuristic methods based on the exact formulation hold promise and are a subject of ongoing research. Such methods have already been developed for the decomposition problem and will be described in the next chapter. Constraint satisfaction methods that target both input and output constraints have also been successfully used to develop heuristic state encoders.

While theoretical groundwork has been laid for the input encoding problem targeting multilevel logic, theory and algorithms for the output and state encoding counterparts are less well-developed. Simple multilevel logic transformations like common cube extraction have been studied from the point of view of prediction at the symbolic level for functions with symbolic inputs and outputs. Heuristic state assignment programs that specifically target multilevel logic have been developed, and produce better multilevel implementations than their two-level counterparts. However, efficient multilevel symbolic output minimization remains a major theoretical and practical challenge in the area of optimal output and state encoding.

Chapter 6

Finite State Machine Decomposition

6.1 Introduction

Finite state machine (FSM) decomposition is concerned with the implementation of a FSM as a set of smaller interacting submachines. Such an implementation is desirable for a number of reasons. A partitioned sequential circuit usually leads to improved performance as a result of a reduction in the longest path between latch inputs and outputs. This fact is particularly true when the individual submachines are implemented as Programmable Logic Arrays (PLAs). It appears that the primary interest in using decomposition tools in industry stems from a need to improve the performance of FSM controllers, which often dictates the required duration of the system clock. FSM decomposition can be applied directly when Programmable Gate Array (PGAs) or Programmable Logic Devices (PLDs) are the target technology. Such technologies are characterized by I/O or gate-limited blocks of logic and latches into which the circuit must be mapped. In many cases, it is desirable for reasons of clock-skew minimization or simplifying the layout to distribute the control logic for a data path in such a manner that the portions of the data path and control that interact closely are placed next to each other. FSM decomposition can also be used for this purpose. Partitioning of the logic implementing the FSM could result in simplified layout constraints resulting in smaller chip area. In PLA-based

FSMs, decomposition has the effect of partitioning the PLA that implements the original FSM into smaller interacting PLAs that implement the individual submachines. In such situations, an area reduction can be attributed to PLA partitioning. Finally, it is not computationally feasible for current multilevel logic minimizers (*e.g.* MIS-II [10]) to search all possible area minimal solutions. In some cases, an initially-decomposed FSM could correspond to a superior starting point for multilevel logic minimization.

It should be noted that performing the decomposition at the State Transition Graph (STG) level where that states are still symbolic, as against partitioning the logic *after* an encoding of states in the original machine and a subsequent logic minimization have already been performed, makes it possible to search a larger solution space for good decompositions. The work presented in this chapter addresses the problem of the decomposition of sequential machines into smaller interacting submachines, so as to optimize for area and achieve the other desirable features like improved performance of the resulting implementation.

The chapter is organized as follows. Basic definitions relevant to this chapter are presented in Section 6.2. We describe approaches to FSM cascade [44] and general [29] decomposition that use the number of states and edges in the resulting submachines as their cost function, in Section 6.3 and Section 6.4, respectively. Given that the logic implementation of a FSM is derived from its STG specification after state assignment and intensive logic optimization, this cost function does not reflect the true complexity of the eventual logic-level implementation and is, on occasion, far from accurate.

We next show that state partitioning can be tied in with the predictive state encoding methods described in Chapter 5 in order to target decompositions with logic-level optimality. A formulation of the optimum two-way decomposition problem targeting two-level logic as one of *symbolic-output partitioning* (originally published in [4]) is presented, in Section 6.5. The cost function associated with this formulation is the total number of product terms in the minimized symbolic representations of the submachines. This cost function is much closer to the final logic-level complexity than the number of states/edges in the decomposition. An exact solution under the formulation chosen above, to the decomposition problem via a method of prime implicant generation and constrained covering is presented. Subsequently, the correctness of the procedure is formally demonstrated. Exact methods for state assignment have been described in Chapter 5, but here the fact that the problem of two-way FSM decomposition is easier than that of state assignment is exploited.

A consequence of the exact decomposition procedure is that two-way or multi-way, parallel, cascade, general or arbitrary decomposition topologies can be targeted, simply by changing the constraints in the covering step. We describe these extensions in Section 6.6. The development of a heuristic optimization strategy, that is applicable to large sized problems is described in Section 6.7. The heuristic procedure consists of an iterative optimization strategy involving *symbolic implicant expansion and reduction*, modified from two-level Boolean minimizers. Reduction and expansion are performed on functions with symbolic, rather than binary-valued outputs. Different expansion/reduction heuristics that have been implemented and evaluated under this global strategy are described.

Section 6.8 explains the relationship between FSM decomposition and state assignment. Experimental results on area and performance optimization are presented in Section 6.9.

6.2 Definitions for Decomposition

A **partition** π on the set of states, S, in a machine is a collection of *disjoint* subsets of states whose set union is S. The disjoint subsets are called the **groups** or **blocks** of π. Consider for example a set of five states, $S = \{s_1, s_2, s_3, s_4, s_5\}$. $\pi = \{(s_1, s_3), (s_2), (s_4, s_5)\}$ is said to be a partition on S. $(s_1, s_3), (s_2)$ and (s_4, s_5) are said to be the blocks of π. Let $\pi_1 = \{a_i\}$ and $\pi_2 = \{b_j\}$, where a_i and b_j are blocks of states, be two partitions on S. The **product** of the partitions π_1 and π_2, denoted $\pi_1 . \pi_2$, is defined as the partition $\pi_1 \cdot pi_2 = \{a_i \cap b_j\}$. The **zero-partition** on S is a partition such that the cardinality of all the groups in it is unity.

Given a State Transition Graph description of a desired terminal behavior, the essence of the decomposition problem is to find two or more machines which, when interconnected in a prescribed way, will display that terminal behavior. The individual machines that make up

$$
\begin{array}{llll}
0 & s1 & s3 & 1 \\
1 & s1 & s1 & 0 \\
0 & s2 & s4 & 0 \\
1 & s2 & s1 & 1 \\
0 & s3 & s2 & 0 \\
1 & s3 & s4 & 0 \\
0 & s4 & s2 & 1 \\
1 & s4 & s3 & 1
\end{array}
$$

Figure 6.1: A STT representation of a Finite State Machine

the overall realization are referred to as **submachines**. Each submachine corresponds to a partition on the set of states, S. All the states belonging to a single block in a submachine are given the same code in that submachine. Therefore, there is no way of distinguishing between two states belonging to a single block in a submachine without recourse to information from other submachines. A block of states in a partition effectively corresponds to a state in the submachine associated with that partition. The **prototype machine** corresponds to the machine that was used to define the terminal behavior to be realized. The term **lumped machine** is used sometimes to denote an undecomposed implementation of the prototype machine. The machine that results as a consequence of the decomposition is called the **decomposed machine**. The functionality of the prototype machine is maintained in the decomposed machine if the partitions associated with the decomposition are such that their product is the zero-partition on S.

Finding the partitions corresponding to the submachines of a decomposition that results in a maximum overall reduction of circuit complexity, is the essence of the optimization problem in machine decomposition. Several methods of finding such partitions for various circuit topologies have been proposed in the past. In the next two sections, we will describe representative methods for the cascade and general decomposition of sequential machines represented by State Transition Graphs.

6.3 Preserved Covers and Partitions

We will begin with an example. Consider the FSM shown in Figure 6.1. Let the two partitions on S be $\pi_1 = \{(s_1, s_2), (s_3, s_4)\}$ and $\pi_2 = \{(s_1, s_3), (s_2, s_4)\}$. The product of these two partitions, $\pi_1 . \pi_2$, is the zero-partition $\{(s_1), (s_2), (s_3), (s_4)\}$. What that means is that given one block of states from each of π_1 and π_2, the state corresponding to the prototype machine is uniquely identified. Now assume that the output logic is implemented only in the submachine corresponding to π_2. In that case, it can be verified from Figure 6.1 that in the submachine corresponding to the partition π_1 (which only implements the next-state logic), there is no need for information from the submachine corresponding to the partition π_2 since the transitions between the blocks in the submachine for π_1 are independent of the block that the submachine for π_2 is in. A partition like π_1 is called a **closed** (or, **preserved**) **partition**. If a closed partition can be found, it means that the prototype machine can be decomposed into a **cascade** of FSMs with the submachine corresponding to the closed partition being the head machine. A **parallel** decomposition exists when closed partitions can be found such that the product of these closed partitions is the zero-partition.

The concept of partitions can be generalized to **covers** [44, 45]. Covers differ from partitions in that blocks in the same cover are allowed to intersect while the blocks in a partition are disjoint. Allowing intersections between blocks in the same cover effectively corresponds to splitting states in the prototype machine without changing functionality and finding partitions on this new set of states. The use of state splitting allows decompositions and state encodings with closed partitions (that would not have been found otherwise) to be found.

Methods for finding preserved covers and partitions have been a subject of investigation [44, 45]. However, systematically searching for all possible decompositions is difficult under the above formulation of the decomposition problem. Further, benchmark studies [34] have indicated that most FSM controllers do not typically have good cascade decompositions. As mentioned in the introduction, a more basic shortcoming of these techniques is that the cost function used, namely the number of states in the decomposed realizations, does not necessarily reflect the eventual logic-level complexity of the realizations. One may pick a preserved cover over another because the former results in a smaller total number of states in the decomposed submachines, but the latter could require smaller total layout area after the decomposed submachines have been encoded and optimized. In some cases, however, these methods used heuristically can be useful as a quick, initial means of finding good decompositions.

6.4   General Decomposition Using Factors

6.4.1   Introduction

General decompositions can have various topologies. The decomposition topology of Figure 6.2 is the one considered in this chapter. In Figure 6.2(b) the original machine, M, has been decomposed into two submachines, M_1 and M_2, interconnected in the prescribed way. The output logic for the decomposed machine is considered to be distributed between the two submachines, or a logic block external to the submachines may be used to generate the primary outputs, as in Figure 6.2.

In this section, we first present a form of general decomposition based on *factorization*. We will define the notion of an exact factor in Section 6.4.3, which results in maximally reducing the number of states and transition edges in the original machine. We will then present a procedure to find all exact factors in a machine in Section 6.4.4.

Exact factors may not exist in a given machine or may be too small to produce a worthwhile decomposition. Inexact but good factors can be found to produce economical decomposed realizations. Techniques to find good, though inexact, factors in a machine have been developed, and will be presented in Section 6.4.5.

6.4.2   An Example Factorization

We will consider the general decomposition topology of Figure 6.2 with an output logic block that corresponds to the OR of the corresponding outputs from machines M_1 and M_2.

An example factorization of a six-state machine is illustrated in Figure 6.3. The original State Transition Graph is given in Figure 6.3(a).

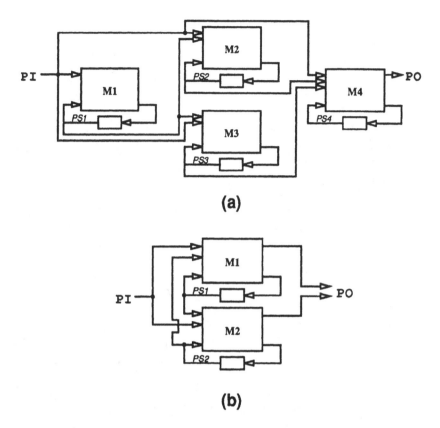

(a)

(b)

Figure 6.2: General decomposition topology

A factor consisting of two states, occurring twice, is identified and extracted. The factoring machine, M_2 is shown in Figure 6.3(c) and the factored machine, M_1, in Figure 6.3(b). Together, they perform the same function as the machine of Figure 6.3(a). The state sets **s2, s3** and **s4, s5** correspond to occurrences of the factor in the original machine. They are replaced by calls to the factoring machine (states **d1** and **d2**). These states are called **calling** states. In M_1 (M_2), the input combination corresponding to each edge has both the primary input vector as well as the state of M_2 (M_1). The loop edges in states **d1** and **d2**, shown in dotted lines, have input combinations equal to the complement of the remaining fanout edges from the corresponding state.

Given a machine, we can identify *any* N_R sets of N_F states as occurrences of a factor. The only restriction is that no state (in

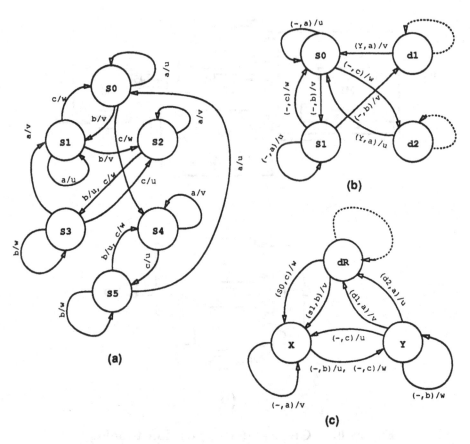

Figure 6.3: Example decomposition

the original machine) can belong to more than one set. If the original machine had N_S states the factored machine M_1 will have $N_S - N_R \times (N_F - 1)$ states and M_2 will have $N_F + 1$ states. However, the complexity of the decomposed submachines, M_1 and M_2, is profoundly affected by the choice of the factor.

The transition edges between the N_R sets of states play an important role in determining the quality of a factor. If exactly similar transition edge relationships exist between these sets of states, factorization will result in the smallest possible number of transition edges in the decomposed submachines. The flow of state information between the two machines, M_1 and M_2, will be minimal and an economical realization will result. However, if the sets of states corresponding to the occurrences of the factor have dissimilar transition edge relationships, the resulting submachines will be complex, *i.e.* with a large number of transition edges dependent on state information from the other machine.

6.4.3 Defining An Exact Factor

We now present definitions which will be used to gauge the quality of a decomposition and which provide a formal basis for the factorization algorithms presented in Sections 6.4.4 and 6.4.5.

A **factor** is N_R (≥ 1) sets of states and all fanout edges from these sets of states in the given machine. Each set of states is called an **occurrence** of the factor F and is denoted $O_F{}^i$. The maximum number of states in any of the N_R occurrences of F, is denoted N_F ($N_F \geq 2$).

A transition edge in the occurrence of a factor, $O_F{}^i$, is an **internal edge** if it fans into and fans out of states within $O_F{}^i$.

An **exit state** in any $O_F{}^i$ is one which has no internal fanout edges.

A **E-reachable factor** is a factor where, in each occurrence $O_F{}^i$, at least one state exists from which all the other states in $O_F{}^i$ can be reached. This state is called the **entry** state of $O_F{}^i$. A E-reachable factor, F, can have more than one entry state and no more than $N_F - 1$ exit states in any occurrence $O_F{}^i$. If unspecified, a factor is assumed to be E-reachable.

A state in $O_F{}^i$ is an **internal factor state** if all edges from the state fan into states within $O_F{}^i$ alone, *i.e.* all fanout edges from the state are internal edges.

Given two occurrences of a factor, $O_F{}^1$ and $O_F{}^2$, a **state correspondence pair** is $(q_1, q_2) \mid q_1 \in O_F{}^1, q_2 \in O_F{}^2$. A factor is defined to be **exact** if:

1. State correspondence pairs for all states in $O_F{}^1$ to states in $O_F{}^2$ can be found such that no state appears in more than one correspondence pair and

2. For each internal edge in $O_F{}^1$, $e1$, if $e1.Input \cap e2.Input \neq \phi$ for any $e2$ in $O_F{}^2$, $(e1.Present, e2.Present)$ and $(e1.Next, e2.Next)$ are state correspondence pairs.

The definition of an exact factor can be extended for $N_R > 2$.

If two edges in a State Transition Graph can be represented by the same product term in an encoded and minimized two-level implementation, these two edges are said to be **mergeable** under that encoding.

6.4.4   Exact Factorization

In this section, we will present an algorithm for finding all exact factors given a STG description of a machine. We also present some theoretical results which provide lower bounds on the number of states and edges in the decomposed submachines after exact factoring.

If a factor is exact or inexact, the numbers of states in the decomposed submachines are only dependent on the number of occurrences of the factor, F, namely N_R and the number of states in the factor N_F. The number of transition edges in the decomposed submachines is, however, related to the similarity between the occurrences of the factor, *i.e.*, whether the factor is exact or inexact.

Theorem 6.4.1 *A decomposed submachine, M_1, produced by factorization via an exact factor with $N_I(i) + N_E(i)$ states in each occurrence $O_{F^i} \in M$, $N_I(i)$ of which are internal states and $N_E(i)$ of which are exit states, will have*

$$\sum_{i=1}^{N_R} (\ |e(i)| - N_I(i) \ )$$

edges less than the original machine, M, where $e(i)$ is the set of internal edges in $O_{F^i} \in M$. The number of edges in the factoring submachine, M_2, will be

$$\leq \ MAX_i \ |e(i)| \ + \ \sum_{i=1}^{N_R} (\ |fin(i)| + |fout(i)| \ )$$

where $fin(i)$ is the set of fanin edges to O_{F^i} and $fout(i)$ is the set of fanout edges from O_{F^i}.

Proof. In the decomposed submachine M_1, the internal edges, $e(i)$ in each O_{F^i} do not exist. The only additional edges in M_1 are the loop edges for each calling state. Since each O_{F^i} has $N_E(i)$ exit states, when M_2 reaches any of these states, control is passed back to M_1. On the other hand, if M_2 is in any of the $N_I(i)$ internal states control remains in M_2 for another clock cycle (at least). $N_I(i)$ loop edges for each calling state will suffice to realize this specification in M_1. The fanout edges from the exit states $fout(i)$ will exist in both M_1 and M. Therefore, the number of edges in M_1 is

$$\sum_{i=1}^{N_R} (\, e(i) - N_I(i) \,)$$

less than the number of edges in M.

From the definition of an exact factor, the internal edges in each $O_F{}^i$ are in one-to-one correspondence, *i.e.* they assert the same outputs if driven by the same input combination. Thus, in the factoring machine, for any call to the factor, these edges can be represented by a single set of edges which are *independent* of the calling state in M_1. A one-to-one correspondence may or may not exist between the edges in $fin(i)$ or $fout(i)$. Therefore, the number of edges in the factored machine, M_2 is:

$$\leq\ MAX_i\ |e(i)|\ +\ \sum_{i=1}^{N_R} (\ |fin(i)| + |fout(i)|\ )$$

and we have the above result. □

Exact factorization can result in significant reductions in the total number of edges and states. It is thus worthwhile to find exact factors in machine, if they exist. We give below a procedure to find all exact factors in a machine given its STG specification. The procedure as presented finds factors which occur twice ($N_R = 2$) but can be generalized for arbitrary N_R.

First, all sets of states of cardinality equal to N_R whose edges assert the same outputs if driven by the same input combination, *regardless of what states they fan out to*, are found. At each (outermost) loop of the algorithm, beginning from a set of N_R states S_{new}, correspondence sets are built based on the fanouts from the set of states (condition **c3** holds). If at any point there is a contradiction in the correspondence sets, *i.e.* condition **c2** holds, the search is aborted and begun with a new starting set, S_{new}. If neither conditions **c1**, **c2** or **c3** hold for a particular set of edges, they are deemed external (not internal). External edges

find-exact-factors:
{

   find all sets of states of cardinality N_R with
   identical I/O fanouts. Store in T_S.

   foreach ($S_I \in T_S$) {
        $S = S_{new} = S_I$; $S_{loop} = \phi$;
        foreach ($Q \in S_{new}$) {
             $q_1 = Q[1]$; $q_2 = Q[2]$;
             foreach ($e_1 \mid e_1.Present = q_1$ &&
                              $e_2 \mid e_2.Present = q_2$) {
                  if ($e_1.Input \cap e_2.Input \neq \phi$) {
                       $n_1 = e_1.Next$; $n_2 = e_2.Next$;
                       c1: if ($(n_1, n_2) \in S$) continue;
                       c2: if ($(n_1 \in S[1]$ && $n_2 \notin S[2])$ ||
                            $(n_1 \notin S[1]$ && $n_2 \in S[2])$)
                            begin next outermost loop iteration ;
                       c3: if ($(n_1, n_2) \in T_S$ || $(n_2, n_1) \in T_S$) {
                            $S = S \cup (n_1, n_2)$;
                            $S_{loop} = S_{loop} \cup (n_1, n_2)$;
                       }
                  }
             }
        }
        if ($|S_{loop}| = 0$ && $|S| > 1$) {
             if (**append-exit-states**(S) == TRUE)
                  S is an exact factor ;
             begin next outermost loop iteration ;
        } **else**
             $S_{new} = S_{loop}$; $S_{loop} = \phi$;
   }
   }
}

Figure 6.4: Procedure to find all exact factors

do not have to satisfy correspondence relationships, but the states they

```
append-exit-states( S ) :
{
    ExitStateSet = φ ;
    foreach ( Q ∈ S ) {
        q₁ = Q[1] ; q₂ = Q[2] ;
        if ( stateFanout(q₁) ∉ S[1] && stateFanout(q₂) ∉ S[2] )
            ExitStateSet = ExitStateSet ∪ Q ;
    }
    foreach ( Q ∈ S | Q ∉ ExitStateSet ) {
        q₁ = Q[1] ; q₂ = Q[2] ;
        foreach ( edge e | e.Present = q₁ ) {
            estate = e.Next ;
            if ( estate ∈ S[1] ) continue ;
            if ( stateFanout(estate) ∉ S[1] )
                ExitStates = ExitStates ∪ S ;
            else return(FALSE) ;
        }
        foreach ( edge e | e.Present = q₂ ) {
            estate = e.Next ;
            if ( estate ∈ S[2] ) continue ;
            if ( stateFanout(estate) ∉ S[2] )
                ExitStatesSet = ExitStatesSet ∪ S ;
            else return(FALSE) ;
        }
    }
    S = S ∪ ExitStatesSet ;
    return(TRUE) ;
}
```

Figure 6.5: Procedure to derive exit states

fan out from have to be exit states in an definition of an exact factor.
This is checked for in the routine **append-exit-states**() which derives
the exit states if the factor is exact. The function $stateFanout(arg)$ in
append-exit-states() returns all states such that an edge from arg to

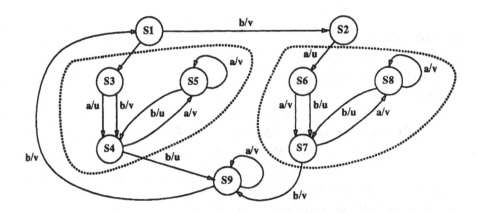

Figure 6.6: Example of detecting exact factors

each of these states exists.

In Figure 6.6, an STG is shown which contains exact factors. We will illustrate how these factors are detected using the procedures described.

1. T_S is constructed. **(s1, s3) (s3, s6), (s4, s7), (s5, s8)**, and **(s7, s9)** are in T_S.

2. We will begin with the state correspondence pair **(s1, s3)**.

3. The fanout on input **a** is **(s3, s4)** which are not in correspondence.

4. We pick the next state correspondence pair **(s3, s6)**. The fanout of this pair on inputs **a** and **b** is the pair **(s4, s7)**. The fanouts are in correspondence.

5. We find that **(s3, s4) (s6, s7)** is an exact factor.

6. We trace the fanout of **(s4, s7)**. The fanout on input **a** is **(s5, s8)**. The fanout on input **b** is **(s9, s9)**.

7. We find that **(s3, s4, s5) (s6, s7, s8)** is an exact factor.

It can be formally proved that the algorithm described finds all exact factors in a STG specification [29].

6.4.5 Identifying Good Factors

Exact factors may not exist in a given State Transition Table specification. However, this does not mean that a good decomposition of the machine cannot be found. The similarity of transition edge relationships existing between the sets of states within each occurrence of the factor $O_F{}^i$, is an indication of the quality of a factor. The number of states in the decomposed submachines M_1 and M_2 is, however, independent of the quality of a factor. The number of edges in the decomposed submachines M_1 and M_2 will be larger for an inexact factor than the values for an exact factor (by Theorem 6.4.1).

To produce good decompositions, factors that are near-exact with maximally similar occurrences have to be identified. A procedure with the same overall structure as the algorithm **find-exact-factors**() can be used.

1. Find similarity weights for all possible N_R sets of states. These weights are found on the basis of I/O fanout relationships between each set of states, *i.e.* the number of input symbols for which edges fanning out of all states in the set have different outputs. A weight of zero would correspond to exactly similar states.

2. These sets are ordered in terms of increasing weights (decreasing similarity). Beginning from each initial set, the fanouts of the set are traced as in **find-exact-factors**().

3. The fanout states from the initial set are found, and depending on their relative similarity weights, a decision is made whether to add them to the factor or not (corresponds to condition **c3** in **find-exact-factors**()). This decision is made based on the cumulative similarity weights of all state sets in the current factor − if the cumulative weight exceeds a certain value, the search is aborted and a new iteration begun.

Thus, given N_R and an upper bound on the desired cumulative similarity weight for a factor, we can find good, though possibly inexact, factors in M.

6.4.6   Limitations of the Factoring Approach

The factoring approach to decomposition suffers from the same limitations that afflict early approaches to decomposition based on preserved covers. While inexact factors will always exist in machines, the cost function used in this approach, namely the number of states and/or edges does not reflect true logic-level complexity. In order to pick a good factor, one needs to measure the quality of the decomposition resulting from the factorization. Accurately gauging the quality of a particular factor requires that we accurately predict the complexity of the encoded and optimized submachines. Comparing state/edge counts does not in general provide an accurate enough measure.

Experimental results provided in [29] indicate that decomposing a machine that has large exact or near-exact factors typically results in a reduction in complexity. Decomposition of large machines into smaller submachines can provide a guide to the state assignment step, making the task of finding an optimal encoding easier. Further, in certain cases reductions in final product term count over one-hot coding (*cf.* Section 3.2) are guaranteed if exact factors are extracted from a machine [29].

6.5   Exact Decomposition Procedure for a Two-Way General Topology

We will now use the notion of generalized prime implicants presented in Chapters 4 and 5 to provide exact or near-exact solutions to the problem of two-way general decomposition targeting two-level logic. To that effect, a formulation of the two-way decomposition problem with an associated cost function that is close to the cost of the ultimate logic-level implementation is presented in this section. Subsequently, an algorithm is presented for finding a decomposition that is optimum under the chosen formulation.

The formulation of optimum decomposition is presented in Section 6.5.2, and its relationship to partition algebra (*cf.* Section 6.3) and factorization (*cf.* Section 6.4) is described in Sections 6.5.3 and 6.5.4, respectively. The various steps in the algorithm are described in Sections 6.5.5 through Section 6.5.10.

6.5.1   The Cost Function

The cost function for a general decomposition can vary depending on the eventual targeted implementation. Here, we are concerned with two-level implementations. The cost function used allows the decomposition of the prototype machine into submachines such that the sum of the areas of the two-level implementations of each submachine, after the states in the submachine have been assigned codes, is less than or close to the area of the two-level implementation of the prototype machine after state assignment. The area of the two-level implementation of each submachine is always less than the area of the two-level implementation of the prototype machine. It is also found empirically that the cost, measured in terms of the area of the standard-cell layout of the logic, of the multilevel implementation of the decomposed machine obtained using this cost function is generally less than the cost of the multilevel implementation of the prototype machine. This implies that an optimal decomposition targeting a two-level implementation is an acceptable decomposition for the multilevel case.

Consider the submachines in Figure 6.2(b). Let the number of product terms in the prototype machine, M, after one-hot coding and two-level Boolean minimization be P. Let the number of product terms in the submachines M_1 and M_2 after one-hot coding and two-level minimization be P_1 and P_2, respectively. A decomposition is deemed to be optimum (optimal) in this formulation if $P_1 + P_2$ is minimum (minimal). In the case when no good decomposition can be found, $P_1 + P_2 = P$.

The area and performance of a PLA are closely related to each other. Since the two-level area of each submachine obtained using this cost function is always less than the two-level area of the prototype machine, and because the critical path of the decomposed machine in the topology of Figure 6.2 is equivalent to the critical path of the larger of the two submachines, the critical path of the decomposed machine in Figure 6.2 will be smaller than the critical path of the prototype machine in the *two-level implementation*. To optimize the critical path of the decomposed machine, the larger of the two submachines should have as small an area as possible. One way of achieving this is to keep the sizes of the submachines similar while minimizing the overall area. Thus, a modified cost function of the form $P_1 + P_2 + \alpha\|P_1 - P_2\|$, in which an additional cost is attached to a difference in the areas of the

PLAs, characterizes the optimality of the decomposition with respect to timing also. α is an empirically determined constant.

6.5.2 Formulation of Optimum Decomposition

The formulation of the optimum decomposition problem is described in the sequel. The first step is to obtain the initial symbolic covers representing the two submachines. To start with, one is given the initial State Transition Table (STT) (or equivalently, the STG) of M. Assume M has N states, s_1, .. s_N. A function L is constructed as follows: The present-state (PS) field in the STT is replaced by an N-valued variable represented by an N-bit vector in which each bit is associated with the presence or the absence of a state. The next-state (NS) field in M is *split* into *two* symbolic variables, *i.e.* s_1 is split into symbolic variables sa_1 and sb_1, s_2 is split into symbolic variables sa_2 and sb_2 and so on. The primary input (PI) and primary output (PO) fields are unchanged. An example transformation is shown below. The example has one primary input, two primary outputs and three states. In the transformed symbolic cover, say L, the first column represents the primary input as before, the second, third and fourth columns correspond to the three-valued variable representing the present state. The fifth and sixth columns represent the partitioned next-state field, and the seventh and eighth columns represent the primary outputs.

$$0 \; s_1 \;\; s_2 \;\; 10 \quad \longrightarrow \quad 0 \; 100 \;\; sa_2 \; sb_2 \; 10$$

$$1 \; s_1 \;\; s_3 \;\; 01 \quad \longrightarrow \quad 1 \; 100 \;\; sa_3 \; sb_3 \; 01$$

$$0 \; s_2 \;\; s_1 \;\; 11 \quad \longrightarrow \quad 0 \; 010 \;\; sa_1 \; sb_1 \; 11$$

$$1 \; s_2 \;\; s_3 \;\; 10 \quad \longrightarrow \quad 1 \; 010 \;\; sa_3 \; sb_3 \; 10$$

$$0 \; s_3 \;\; s_2 \;\; 00 \quad \longrightarrow \quad 0 \; 001 \;\; sa_2 \; sb_2 \; 00$$

$$1 \; s_3 \;\; s_1 \;\; 01 \quad \longrightarrow \quad 1 \; 001 \;\; sa_1 \; sb_1 \; 01$$

The next step is to obtain the symbolic covers for the submachines. The symbolic cover, L_a, for **Submachine a** is obtained in the following manner. For each row in L, there is a corresponding row in L_a that has the same PI and PS fields as in L. The next-state field in a row in L_a consists of the symbol sa_i from the partitioned next-state field in the corresponding row in L. The symbolic cover, L_b, for **Submachine b** is obtained in a similar manner, with the next-state field in a row of L_b being the symbol sb_i from the corresponding row in L. The primary outputs in L are partitioned between L_a and L_b. By *partitioning* of the primary outputs, it is implied that some of the primary outputs are now asserted in the symbolic cover for **Submachine a** while the remaining primary outputs are asserted in the symbolic cover for **Submachine b**. In the case when the primary outputs are also symbolic and need to be encoded, they can also be symbolically partitioned and treated in the same manner as the next-state lines. In the example shown below, all the primary outputs in L are asserted in L_a (shown on the left) and no primary outputs are asserted in L_b (shown on the right).

0 100	sa_2	10	
1 100	sa_3	01	
0 010	sa_1	11	
1 010	sa_3	10	
0 001	sa_2	00	
1 001	sa_1	01	

L_a

0 100	sb_2
1 100	sb_3
0 010	sb_1
1 010	sb_3
0 001	sb_2
1 001	sb_1

L_b

L_a and L_b are identical to each other and to L, with the difference that some of the fields in the next-state plane and output plane of L are asserted in L_a while the remaining are asserted in L_b. L_a and L_b represent the initial symbolic covers of the two component submachines that M is to be decomposed into. The column in L_a with the sa_i's represents the next-state column for **Submachine a**, and the column in L_b with the sb_i's represents the next-state column for **Submachine b**. The operation of **Submachine a** and **Submachine b** is mutually dependent, with the communication between the submachines, as can be seen in Figure 6.2(b), being via the present-state lines of each submachine. While the next-state field was partitioned in the above procedure, the present-state field, represented by the multiple-valued (MV) variable, was duplicated in L_a and L_b. As in L, the MV input variable in L_a and L_b also represents states in the *prototype machine*. Since the state of

Submachine a and the state of **Submachine b** together uniquely determine the state of the overall machine M and since a knowledge of the state of the overall machine is sufficient to determine the states of **Submachine a** and **Submachine b**, the MV input variable in L_a and in L_b represents, implicitly, the present-state of **Submachine a** as well as of **Submachine b**.

The symbolic outputs in L_a and L_b are now represented by means of a **one-hot code** (*cf.* Section 3.2 and Section 5.1.1). Given the one-hot coded initial symbolic covers for the two submachines, the goal is to minimize the sum of the cardinalities of the two covers. To achieve this goal, one can use the degree of freedom that since each submachine knows its own present-state and the present-state of the other submachine, as long as two states of the prototype machine have been given different codes in *one* of the submachines, it is possible to distinguish between the two states in the decomposed machine. For example, states s_1 and s_2 can be given the same code in **Submachine a**, *i.e.* $enc(sa_1) = enc(sa_2)$, as long as it is possible to distinguish between the two states by means of the codes given to them in the other submachine. Since the encoding is one-hot, two states, say sa_i and sa_j or sb_i and sb_j, either have the same code or the bitwise intersection of their codes is null. The symbolic output, in this case the next-state variable, is represented by sets, called **tags**, of next-state symbols that have been given the same code. Initially, when all states have different codes, the symbolic-output tag for a cube consists of only a single next-state symbol corresponding to the next-state asserted by the cube. When two states are given the same code, the symbolic-output tags containing these two states are replaced by tags that contain both the next-state symbols. For example, when sa_i and sa_j are given the same code, each occurrence of the symbolic-output tags (sa_i) and (sa_j) is replaced by the tag $(sa_i\ sa_j)$. A similar replacement is carried out when three states are given the same codes, and so on.

Let the final cardinality of the minimized cover for **Submachine a** be P_a and that for **Submachine b** be P_b. Let the cardinality of the minimized cover for the prototype machine, with the states one-hot coded, be P. Consider for illustration the extreme case where all the sa_i are given the same code. To distinguish between the states of the prototype machine in this case, all the sb_i have to be assigned distinct codes. Effectively, **Submachine a** realizes the primary output logic of the prototype machine and **Submachine b** realizes the next-state logic and as a result $P_a + P_b \geq P$. In the other extreme case, if all the sb_i are given the same code, then **Submachine b** is not required at all (*i.e.* $P_b = 0$) since **Submachine b** does not assert any primary outputs, but all the sa_i have to be given different codes and hence $P_a = P$.

6.5.3 Relationship to Partition Algebra

The problem, therefore, is to decide which states of the prototype machine should be given the same code and the submachine in which they should be given the same code. The states that are given the same code in a submachine belong to the same block in the *partition* (*cf.* Section 6.2) corresponding to that submachine. This problem is identical to finding two or more partitions such that the sum of the cardinalities of the minimized symbolic covers of the submachines represented by these partitions is minimum and the product of the partitions is the zero-partition.

A large number of choices are available in the states that can be given the same codes. In order to avoid enumerating all the large number of possibilities of states being given the same codes, a formal procedure is required that searches all the possibilities exhaustively, but implicitly.

6.5.4 Relationship to Factorization

It was suggested in Section 6.4 that typical FSMs possess isomorphic or close to isomorphic subgraphs in their STGs. Identifying such isomorphic subgraphs and implementing multiple instances of an isomorphic subgraph as a single separate submachine distinct from the parent machine corresponds to a method for decomposition. The effect of such a decomposition is to identify instances of similar functionality in the STG and to encapsulate them into a single submachine, or in other words identify subgraphs that it is advantageous to replace with subroutine calls. This *factorization* approach is equivalent to identifying specific types of partitions on the states in the prototype machine. In the ideal situation, the procedure for decomposition proposed in this chapter should automatically be able to identify such partitions if they

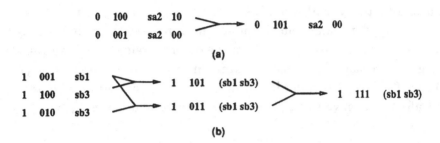

Figure 6.7: Examples of cube merging

translate into lower logic-level complexity.

6.5.5   Generalized Prime Implicant Generation

To solve an output encoding problem, one has to modify the prime implicant generation and covering strategies basic to Boolean minimization (*cf.* Section 5.2 of Chapter 5). The decomposition problem is slightly different and simpler than the classical output encoding problem since a one-hot coding has already been assumed, and the only degree of freedom is in giving the same code to the symbolic outputs.

Prime implicants and covering are basic to two-level Boolean minimization because of the fact that a minimum/minimal cardinality two-level cover can always be made up of only prime implicants. In this manner, the two-level Boolean optimization problem can be reduced to a covering problem [68] which involves the selection of an optimum set of prime implicants such that the functionality of the original specification is maintained. It is shown in the sequel that it is possible to define a notion of primality in terms of generalized prime implicants (GPIs) so that the decomposition of FSMs can also be formulated as a covering problem on GPIs. As in two-level Boolean minimization and output encoding, it can be shown in the case of FSM decomposition that a minimum cardinality solution can be made up exclusively of GPIs. By restricting the search for the optimum solution to sets of GPIs, the explicit, exhaustive enumeration of all possible combinations of cubes is avoided.

Given a symbolic cover with a multiple-valued (MV) input variable and a symbolic output variable in addition to Boolean primary in-

puts and outputs, a GPI is defined as a cube that is not covered by a
larger cube that has the same MV input literal or a MV input literal with
a one in every position, and the same symbolic-output tag.

The procedure for generation of GPIs is similar to the Quine-
McCluskey procedure with the additional tags corresponding to the symbolic output. Initially, all minterms have tags corresponding to the
next-state they assert. Larger cubes are generated by merging smaller
cubes. It is possible for two minterms with the same symbolic output
tag to merge to form a larger cube. An example of two minterms (belonging to the cover L_a shown above), with the same symbolic-output
tag, merging together is shown in Figure 6.7(a). Cubes with different
symbolic-output tags can also merge to form larger cubes. If a cube
that asserts a symbolic-output sa_i, merges with a cube asserting the
symbolic output sa_j, the symbolic-output tag of the resulting cube has
both symbols sa_i and sa_j. An example of this form of merging is shown
in Figure 6.7(b). After larger cubes are generated in this manner, cubes
that are not GPIs can be removed from the set. When no larger cube
can be generated, one has the set of all GPIs. In the example in Figure 6.7(a), the larger cube is not allowed to remove the cubes that it
was obtained from because the MV input literal of the larger cube is not
the same as that of the smaller cubes. In the case of the example of Figure 6.7(b), the cubes **1 011** (sb_1 sb_3) and **1 101** (sb_1 sb_3) are removed
from the selection because the cube **1 111** (sb_1 sb_3) covers them and
has a '1' in every position of its MV input literal.

GPIs are generated in this manner for the two covers, L_a and
L_b, separately. Given the GPIs for L_a and for L_b, the next step is the
selection of a subset of GPIs that covers all the minterms in the initial
symbolic covers for the two submachines such that the cardinality of the
selected set is minimum. Subsequently, the states in the submachines
are encoded according to the choices regarding the states that are to
be given the same/different codes made during selection of GPIs. The
code for a state of the overall machine is then the concatenation of the
codes of the corresponding states in the submachines. For example, the
code for the state s_1 is the concatenation of the codes of the states sa_1
and sb_1. Once the states have been encoded, the MV input literal of
each GPI is replaced by the super-cube containing the minterms corresponding to all the states of the prototype machine present in it. In

addition, the symbolic-output tag of each GPI is replaced by the code given to the states present in it. Once a GPI has been so modified, it represents a cube in the two-level Boolean cover for the submachine to which it belongs. A selection of GPIs is said to be **encodeable** if, after the states of the submachines have been encoded and the GPIs appropriately modified, the functionality represented by the resulting logic is the same as that of the prototype machine. An encodeable, minimum cardinality selection of GPIs represents an upper bound on the number of product terms required in a two-level Boolean cover realizing the decomposed submachine. For reasons explained in the sequel, an arbitrary selection of GPIs that covers all the minterms in the initial covers of the two submachines is not necessarily encodeable. The covering problem that is to be solved for finding the optimum decomposition is, therefore, different from and more difficult than the covering problem associated with classical Boolean minimization. Since the covering procedure is constrained to those selections of GPIs that are legal, it is termed the *constrained covering problem.*

It remains to clearly define how to constrain the selection of GPIs in order to obtain encodeable covers. To that end, the reasons that particular selections may not be encodeable are explained.

6.5.6 Encodeability of a GPI Cover

If the selection of GPIs is such that the two states of the prototype machine are not given a different code in some submachine, it is impossible to distinguish between the two states in the decomposed machine. If the two states are not equivalent, the functionality of the decomposed machine obtained in this manner is bound to be different from that of the prototype machine. Thus, a selection of GPIs that results in the codes for some pair of states to be the same in all submachines is not encodeable. Associated with this reason for the unencodeability of a selection of GPIs is a constraint, termed the *output constraint*, that any encodeable selection must satisfy, which is that the selection of GPIs should allow any pair of states to have different codes in some submachine.

There is a second reason why a selection of GPIs may not be encodeable. As has been stated earlier, the code for a state in the overall machine is a concatenation of the codes of the corresponding states in the

1	001	(sa1 sa2)	01
0	011	(sa1 sa2)	00
0	010	(sa1 sa2)	11
1	010	(sa3)	10
1	100	(sa3)	01
0	110	(sa1 sa2)	10

1	111	(sb1 sb3)
–	010	(sb1 sb3)
0	101	(sb2)

Submachine 'a'   Submachine 'b'

Figure 6.8: An example of an input constraint violation

submachines, *e.g.* the code for state s_1 is a concatenation of the codes for the states sa_1 and sb_1. It has also been stated that, after the encoding of states, the MV input literal of each GPI is replaced by a super-cube that contains the minterms corresponding to the codes assigned to all the states of the prototype machine present in that literal. A condition that the resulting cube must satisfy in order not to alter the functionality of the prototype machine, is that the super-cube mentioned must not contain the codes of the states that are absent from the MV input literal. If all states had been given different codes in each submachine, then there would always exist some encoding that would satisfy this requirement. But, in case some states of the prototype machine are given the same code in a submachine, it is possible that there may exist a GPI for which no encoding of states would be able to satisfy the requirement. The constraint on the selection of GPIs associated with this requirement is termed an *input constraint*. While the input constraint encountered here is similar to that described in Chapter 3, there are some differences. Input constraints in decomposition are best explained by means of an example.

Consider the example shown in Figure 6.8. The figure shows a selection of GPIs obtained from the initial covers L_a and L_b (*cf.* Section 6.5.2). It can be verified that all the minterms in L_a and in L_b are covered by this selection. It can also be verified from the figure that no two states of the prototype machine have been given the same code in all the submachines. All the same, this selection of GPIs is not encodeable because of the presence of the GPI **0 011** (sa_1 sa_2) **00** in the cover for **Submachine a**. The MV input literal of this GPI is **011**. No encoding of the states sa_i and sb_i (and hence of the states s_i) can realize a super-cube that contains the codes of the states s_2 and s_3 but not the code of the state s_1. This is the case because $enc(sa_1) = enc(sa_2)$ and $enc(sb_1) = enc(sb_3)$. Therefore, the code for s_1 is given by $enc(s_1) = (enc(sa_1)$ @ $enc(sb_1)) = (enc(sa_2)$ @ $enc(sb_3))$. [1] Since the super-cube correspond-ing to the MV input literal **011** contains the codes for both the states s_2 $(enc(sa_2)$ @ $enc(sb_2))$ and s_3 $(enc(sa_3)$ @ $enc(sb_3))$, it also contains the codes $(enc(sa_2)$ @ $enc(sb_3))$ and $(enc(sa_3)$ @ $e(sb_2))$, and therefore the code for the state s_1 perforce. This implies that the minterm cor-responding to the code for the state s_1 appears in a cube that it is not supposed to occur in, resulting in the corruption of the fanout edges of the state s_1. This in turn results in the functionality of the decomposed machine not reflecting the functionality of the prototype machine. Since the input constraint is associated with the absence of a state from the MV input literal, the MV input literal in which all states of the prototype machine are present does not represent a constraint.

6.5.7   Correctness of the Exact Algorithm

In this section, it is shown that the procedure formulated above does indeed realize the minimum cardinality solution to the decomposi-tion problem.

Lemma 6.5.1 *Checking for the input and output constraints stated in the previous section is necessary and sufficient to ensure that the func-tionality of the decomposed machine is identical to that of the prototype machine.*

Proof. Necessity: If two states, s_i and s_j of the prototype machine are not given different codes in some submachine, the decomposed machine is unable to distinguish between them. Hence, the functionality is not maintained unless the states s_i and s_j are equivalent. If the states are equivalent, they are not constrained to have different codes. In addition, if the input constraint is violated as illustrated by the example in the previous section, the fanout edges of at least one of the states get corrupted and consequently the functionality of the resulting logic does not reflect the functionality of the prototype machine.

---

[1] The @ operator, as in (i @ j) denotes the concatenation of two strings, in this case i and j.

Sufficiency: To exhibit the sufficiency of the encodeability constraints, it is shown that the functionality of the decomposed machine is identical to the functionality of the prototype machine if the encodeability constraints are satisfied. Since the present-state fields of the prototype machine were not split while obtaining L_a and L_b (*cf.* Section 6.5.2), there is a one-to-one correspondence between the present-state fields of the decomposed machine and the present-state fields of the prototype machine. No fanout edge of any state is corrupted since all the input constraints are satisfied. Also, the selection of GPIs is such that all the minterms of L_a and L_b are covered. Therefore, the functionality of the submachines considered *separately* is maintained. Since a primary output is asserted either in L_a or in L_b, the functionality of the prototype machine with respect to the primary outputs is also retained in the decomposed machine. Also, since the selection of GPIs is such that no two non-equivalent states are assigned the same codes in both the submachines, each pair of next states asserted in the individual submachines for the same primary input and present state combination is associated with a unique next-state in the prototype machine. Because the functionality of L_a and L_b, considered individually, is maintained, this unique next-state is the same as the next-state that would have been produced by the prototype machine for the given primary input and present-state combination. Hence, the functionality of the prototype machine with respect to the next-state logic is also maintained. □

Lemma 6.5.2 *A minimum cardinality encodeable solution can be made up entirely of GPIs.*

Proof. The proof is by contradiction. Assume that one has a minimum cardinality solution with a cube c_1 that is not a GPI. It is known that there exists a GPI covering c_1 that has the same tag as c_1, that its multiple-valued (MV) input part is either the same as that of c_1 or has a 1 in all its positions, that its binary-input part covers the binary-input part of c_1 and its binary-output part covers the binary-output of c_1. If no such GPI exists, then by the definition of GPIs (*cf.* Section 6.5.5) c_1 itself is a GPI. If such a GPI does exist, replacing c_1 by the GPI (**1**) does not change the functionality because all the minterms covered by c_1 are also covered by the GPI, (**2**) does not change the cardinality

```
0 s1 s2 1
1 s1 s3 1
0 s2 s3 0
1 s2 s4 0                                        0 1000 (sb2)      1
0 s3 s3 0     - 1000 (sa1 sa2 sa3)               1 1000 (sb3 sb4) 1
1 s3 s4 0     0 1111 (sa1 sa2 sa3)               - 0110 (sb3 sb4) 0
0 s4 s2 1     - 0001 (sa1 sa2 sa3)               0 0001 (sb2)      1
1 s4 s1 1     1 0110 (s4a)                       1 0001 (sb1)      1

Prototype M/C      Submachine 'a'                 Submachine 'b'
```

Figure 6.9: Example of a general decomposition

because c_1 is being *replaced* by the GPI, and (3) does not change the encodeability of the solution because the symbolic-output tag of c_1 is the same as that of the GPI and the MV input literal of the GPI is either the same as that of c_1 or is all 1's, thus requiring no new encodeability constraints to be satisfied (see Section 6.5.9 for the explanation of why a MV input literal with all 1's does not represent any constraint). Hence, a minimum cardinality solution can be made up entirely of GPIs. □

Theorem 6.5.1 *The selection of a minimum cardinality encodeable GPI cover for L_a and L_b represents an exact solution, that is a solution with minimum cost, to the decomposition problem under the chosen formulation.*

Proof. The proof follows from Lemmas 3.1 and 3.2. □

6.5.8   Checking for Output Constraint Violations

In this section and in Section 6.5.9, we will present an algorithm that can quickly determine the encodeability of a set of generalized prime implicants (GPIs). We will concentrate on output constraint violations in this section, and input constraint violations in Section 6.5.9.

The example of the four-state FSM shown in Figure 6.9 is used to explain this procedure. Say that the selection of GPIs shown in the covers for the submachines in Figure 6.9 has been made for this FSM. It is now required to determine whether this selection of GPIs is encodeable.

To check for output constraint violations, a graph, termed the encodeability graph, is constructed in which each vertex is associated with a state s_i in the prototype machine and an edge occurs with label a from vertex s_i to vertex s_j in the graph if the states s_i and s_j co-exist in the tag of some GPI in **Submachine a**. Similarly, there is an edge with label b from vertex s_i to vertex s_j if the states s_i and s_j co-exist in the tag of some GPI in **Submachine b**. The encodeability graph corresponding to the selection of GPIs in Figure 6.9 is shown in Figure 6.10. The states s_1, s_2 and s_3 occur in the same output tag in the cover of **Submachine a**. Hence, there exists an edge between vertices s_1 and s_2, between vertices s_1 and s_3 and between vertices s_2 and s_3, all with the label a. Similarly, since states s_3 and s_4 occur in the same output tag in the cover for **Submachine b**, there is an edge between the vertices s_3 and s_4 with label b in the encodeability check graph. If any s_i, s_j pair has edges with both labels a and b, it implies that the two states have been assigned the same code in both submachines and the selection is invalid. The dotted edge with label b between states s_1 and s_2 in Figure 6.10 would have existed and caused a constraint violation if the states s_1 and s_2 had been assigned the same code in **Submachine b**.

Since the attempt is to identify partitions in the prototype machine and since groups of states in a partition are disjoint, a transitivity constraint is imposed on the encodeability graph whereby if vertices s_i and s_j have an edge with label a between them and vertices s_j and s_k also have an edge with label a between them, then vertices s_i and s_k must also have an edge with label a between them. In terms of codes given to states, this simply means that if states s_i and s_j are given the same code and s_j and s_k are given the same code in **Submachine a**, then s_i and s_k also should be given the same code in **Submachine a**.

A **clique** is defined here as a subgraph such that each pair of its constituent vertices is connected by edges with the *same* label. Figure 6.11 shows an encodeability graph for a decomposition in which states s_1 and s_2 occur in the same output tag of **Submachine a**. If a new edge with label a is added between vertices s_1 and s_3, it becomes necessary to also add another edge with label a between vertices s_2 and s_3 because of the transitivity requirement. As a result, a clique with label a consisting of vertices s_1, s_2 and s_3 is formed. This implies that the states s_1, s_2 and s_3 all have to occur in the same output tag in **Submachine a**. A state that is not given the same code as any other state forms a single-vertex clique. The encodeability graph is thus composed of a set of cliques satisfying the following properties if the selection of GPIs does not violate an output constraint:

- All the edges in a particular clique have only one type of label. Thus, a clique can be identified with a label. The vertices s_1, s_2 and s_3 form a clique with label a in the graph of Figure 6.10.

- Two cliques with the same label cannot have a vertex in common unless both the cliques are contained in a single large clique. Let a clique with vertices s_i, s_j and s_k and label a be denoted by $(s_i\ s_j\ s_k)_a$. The cliques $(s_1\ s_2)_a$ and $(s_2\ s_3)_a$ in Figure 6.10 have a vertex in common and have the same label, but, are contained in the larger clique $(s_1\ s_2\ s_3)_a$. This property stems from the transitivity requirement.

- Any two cliques can have, at most, one vertex in common. The cliques $(s_1\ s_2\ s_3)_a$ and $(s_1\ s_3)_b$ in Figure 6.10 have different labels and only the vertex s_3 in common. This property follows from the fact the no two states can be given the same code in all the submachines.

To check for an output constraint violation, only a single pass needs to be made through the set of *selected GPIs*. For the output tag of each GPI encountered, the corresponding clique is constructed in the encodeability graph and it is checked if any of the three properties above are not satisfied. Checking for the satisfaction of the properties requires a constant number of Boolean operations, where the complexity of each Boolean operation is of the order of the number of states in the prototype machine. If the properties are satisfied, the clique is added to the encodeability graph. Otherwise the selection of GPIs is not encodeable. Therefore, the complexity of checking for output constraint violations in the encodeability check algorithm is some constant times the product of the number of *selected GPIs* and some constant power of the number of states.

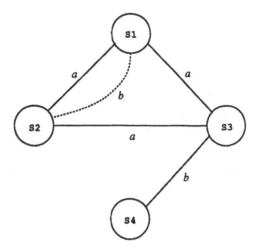

Figure 6.10: Encodeability check graph for the example decomposition

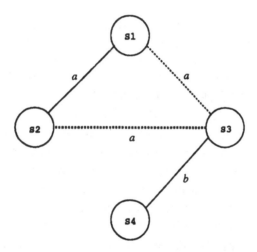

Figure 6.11: Adding a new edge

6.5.9 Checking for Input Constraint Violations

Once it has been verified that the selection of GPIs does not violate output constraints, violations of constraints imposed by the MV-input literals of the GPIs (*cf.* Section 6.5.6) are checked for. Multiple-valued (MV) input literals with a 1 in all the positions represent input constraints that are trivially satisfied. Similarly, those MV input literals with only a single 1 also do not represent constraints because if a super-cube contains only a single minterm corresponding to the code for the state present in the MV literal, one is guaranteed that the code will be different from the code for any other state since no output constraints have been violated. This implies that the super-cube will not contain the code of a state absent from the MV literal.

A selection of GPIs violates an input constraint *if and only if* there exists a MV input literal in one of the selected GPIs and a pair of cliques (*cf.* Section 6.5.8) in the encodeability graph such that the following conditions are satisfied:

- The intersection of the two cliques is non-null.

- The intersection of the two cliques, *i.e.* the vertex common to the two cliques, is a state s_i such that s_i is absent from the MV input literal.

- There is at least one state in *each* of the two cliques that is also present in the MV input literal.

It can be verified that these three requirements follow directly from the basic reason for input constraint violations (*cf.* Section 6.5.6).

If the number of GPIs in the *selected set* is G, the number of cliques with label a is C_a and the number of cliques with label b is C_b, the number of checks required for input constraint violations is $G \cdot C_a \cdot C_b$ in the worst case since each pair of cliques may have to be checked for every GPI. Assume that there are N states in the prototype machine. If a clique is represented by an N-bit vector, like a MV input literal (*cf.* Section 6.5.2), in which a 1 in a position corresponds to the presence of a vertex in the clique, a total of βN (β is a constant) bitwise intersections are required for each of the $G \cdot C_a \cdot C_b$ checks. Therefore, the complexity of checking for input constraint violations is of the order of $\beta \cdot N \cdot G \cdot C_a \cdot C_b$. We know that C_a and C_b are always some fraction of G, and given the initial FSM specification, the number of states, N, is a constant. The complexity of checking for input constraint violations is therefore of the order of some constant power of G.

The procedure for checking for input constraint violations is speeded up significantly by the use of the following pruning techniques that significantly reduce the search space:

- Pairs of cliques that have a null intersection need not be considered.

- Any GPI with a MV input literal in which only a single symbol is present need not be considered.

- Any GPI with a MV input literal in which all symbols are present need not be considered.

- Any GPI with a MV input literal that is contained within some clique need not be considered.

- A GPI-clique pair need not be considered if the clique is disjoint from the MV input literal of the GPI.

- A GPI-clique pair need not be considered if the clique is contained within the MV input literal of the GPI.

Consider the example given in Figure 6.9. The GPIs **- 1000** $(sa_1\ sa_2\ sa_3)$, **- 0001** $(sa_1\ sa_2\ sa_3)$, **0 1000** (sb_2) **1**, **1 1000** $(sb_3\ sb_4)$ **1**, **0 0001** (sb_2) **1** and **1 0001** (sb_1) **1** need not be considered for input constraint violation because all of them have only one state in their MV input literals. The GPI **0 1111** $(sa_1\ sa_2\ sa_3)$ has a MV input literal with all the symbols present and therefore cannot cause an input constraint violation. The cliques (*cf.* Section 6.5.8 for notation on cliques) in the encodeability graph of Figure 6.10 are the following: $(s_1\ s_2\ s_3)_a$, $(s_4)_a$, $(s_3\ s_4)_b$, $(s_1)_b$ and $(s_2)_b$. The single-vertex cliques $(s_1)_b$ and $(s_2)_b$ cannot violate constraints, and are therefore not considered. Thus, a constraint violation can only occur due to the cliques $(s_1\ s_2\ s_3)_a$ and $(s_3\ s_4)_b$. The intersection of these two cliques is the state s_3. Therefore, it is only necessary to consider those GPIs for possibility of violation that have a MV input part with the state s_3 *absent*. None of the remaining GPIs satisfy that requirement. Therefore, no violation of input constraints exists in the selected set of GPIs shown in Figure 6.9.

If neither the output constraint nor the MV input constraint are violated, the cover is deemed encodeable.

6.5.10 Relative Complexity of Encodeability Checking

As has been shown in Sections 6.5.8 and 6.5.9, the complexity of checking for output and input constraint violations is polynomial

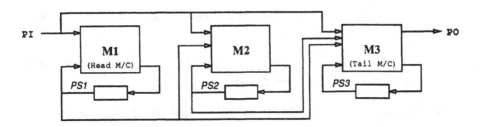

Figure 6.12: Topology for three-way cascade decomposition

in the number of *selected GPIs*. Covering on the other hand is NP-complete in the total number of GPIs [35]. The bottle-neck in the exact decomposition procedure is therefore the covering procedure rather than encodeability checking.

The exact decomposition algorithm can be extended to the problem of decomposition into multiple component machines and in general to obtain decompositions with arbitrary topologies. The reasons that this exact algorithm may not be viable for a given problem are that the number of GPIs may be too large and/or the covering problem may not be solvable in reasonable time. Therefore, a heuristic procedure is required to solve large problems (*cf.* Section 6.7).

6.5.11   The Covering Step in Exact Decomposition

Covering satisfying encodeability constraints is carried out in a manner similar to the procedure described in Section 4.2.9. The lower bounding methods described there can be modified to the encodeability definition for the decomposition case.

6.6   Targeting Arbitrary Topologies

In this section, it is shown that by means of the formulation for decomposition via constrained covering described so far, it is possible to target the logic-level optimality of arbitrary topologies by merely changing the encodeability constraints.

6.6.1 Cascade Decompositions

The topology for a three-way cascade decomposition is shown in Figure 6.12. The characteristic property of cascade decompositions is that information flows in only one direction. For example, a submachine in Figure 6.12 receives as input only the primary input and the present state of the submachines to its left. The submachine that receives as input only the primary inputs is known as the head machine, and the submachine that receives as input the present states of all the submachines is known as the tail machine. Given a prototype machine, one can target such a topology by specifying the appropriate constraints on the selection of GPIs. The initial partitioning along the next-state and primary output fields, and the subsequent generation of GPIs is carried out in the same manner as in the case of general decomposition (*cf.* Section 6.5.2). Imposing the following constraints on the covering procedure then ensures that the topology corresponds to a cascade decomposition:

1. The code of a state in the prototype machine should be different from the code of any other state in at least one of the submachines. A selection of GPIs in which two states are in the same output tag in all the submachines is not encodeable.

2. Say that **Submachine a** is a part of the cascade chain with submachines preceding it and following it. If two states that have not been given different codes in any submachine preceding **Submachine a** occur in the same symbolic-output tag in **Submachine a**, it is not possible for **Submachine a** to distinguish between the two states. Say that a selection of GPIs has been made such that s_1 and s_2 occur in the same symbolic-output tag in **Submachine a**. The selection is valid only if other GPIs for **Submachine a** exist in the selection such that all the pairs of states that are the next states of s_1 and s_2 for each primary input also occur in the same symbolic-output tags in these GPIs. This corresponds to the requirement of a *preserved partition* (*cf.* Section 6.3) in a cascade decomposition. In the case of the head machine, this constraint has to be satisfied for any pair of states that occur in the same symbolic output tag.

```
0 s1 s3 1
1 s1 s1 0
0 s2 s4 0                                          0 1000 (sb1 sb3) 1
1 s2 s1 1                                          1 1000 (sb1 sb3) 0
0 s3 s2 0        1 1100 (sa1 sa2)                  1 0101 (sb1 sb3) 1
1 s3 s4 0        0 0011 (sa1 sa2)                  0 0100 (sb2 sb4) 0
0 s4 s2 1        0 1100 (sa3 sa4)                  - 0010 (sb2 sb4) 0
1 s4 s3 1        1 0011 (sa3 sa4)                  0 0001 (sb2 sb4) 1

Prototype M/C       Head M/C                          Tail M/C
```

Figure 6.13: An example of a two-way cascade decomposition

3. The constraint on the MV input part of the GPIs is slightly different from the case of general decomposition. Again, consider a submachine, say **Submachine a**, that is a part of the cascade chain with submachines preceding it and following it. Let MV_a be a MV input literal of a GPI for **Submachine a**. If a state of the prototype machine is absent from MV_a, the constraint imposed is that the code of the state has to be different from the codes of the states present in MV_a in some submachine *and* that this submachine can only be either **Submachine a** or any of the submachines preceding **Submachine a**.

An example of a two-way cascade decomposition that does not violate any constraint is shown in Figure 6.13. A selection of GPIs that constitutes the cascade decomposition is shown in the figure. It can be seen from the figure that in the head machine, the states that have been given the same codes are contained in the same MV input literal and, consequently, have the same next-state fields. Thus, the second constraint is not violated. It can be verified that the second constraint would have been violated, for example, had states s_1 and s_2 been given the same code in the head machine but states s_3 and s_4 had been given different codes. It can also be verified that the third constraint is not violated. It would have been violated, for example, had the GPI **0 0110** (sb_2 sb_4) **0** been present in the cover for the tail machine because the state s_1, which is absent from the MV input literal of this GPI, does not have a code different from the codes of states s_2 and s_3 in the same submachine.

As in the case of general decomposition (*cf.* Section 6.5.7), it can be shown that the above constraints for decomposition into cascade

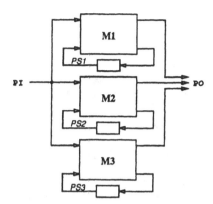

Figure 6.14: Topology for three-way parallel decomposition

chains are necessary and sufficient to ensure that the decomposition has the same functionality as the prototype machine.

Lemma 6.6.1 *The three constraints described in the above section are necessary and sufficient to ensure that a selection of GPIs satisfying them results in a cascade decomposition that has the same functionality as the prototype machine.*

Proof. The proof is similar to the proof for Lemma 6.5.1. □

6.6.2   Parallel Decompositions

The topology for a three-way parallel decomposition is shown in Figure 6.14. The characteristic property of parallel decompositions is that all the submachines operate independently of each other. Thus, a submachine does not receive as input the present-state lines of *any* other submachine. All the submachines are, therefore, similar to the head machine of a cascade chain. Given a prototype machine, one can target such a topology by specifying the appropriate constraints. The initial partitioning and the subsequent generation of GPIs for a parallel decomposition is carried out in the same manner as in the case of general decomposition (*cf.* Section 6.5.2). The following constraints are then imposed on the selection of the GPIs to ensure that a parallel decomposition is obtained:

```
0  s1  s2  01
1  s1  s3  10
0  s2  s1  00
1  s2  s4  11
0  s3  s4  11      0  1100  (sa1 sa2)  0      0  1010  (sb2 sb4)  1
1  s3  s1  00      1  1100  (sa3 sa4)  1      1  1010  (sb1 sb3)  0
0  s4  s3  10      0  0011  (sa3 sa4)  1      0  0101  (sb1 sb3)  0
1  s4  s2  01      1  0011  (sa1 sa2)  0      1  0101  (sb2 sb4)  1

Prototype M/C        Submachine 'a'           Submachine 'b'
```

Figure 6.15: An example of a two-way parallel decomposition

1. The code for a state in the prototype machine should be different from the code for any other state in at least one of the submachines. A selection of GPIs in which two states are in the same output tag in all the submachines is not encodeable. This constraint is identical to the corresponding constraint for general and cascade decompositions.

2. Since no submachine in a parallel decomposition receives as input the present state of any of the other submachines, if the states s_1 and s_2 are given the same code in a submachine, a selection of GPIs is valid only if two conditions are satisfied. Firstly, other GPIs for that submachine exist in the selection such that all the pairs of states that are the next states of s_1 and s_2 for each primary input combination also occur in the same symbolic-output tags. Secondly, the states s_1 and s_2 have to assert the same primary output lines in that submachine for any primary input combination.

3. The third constraint is similar to the corresponding constraint in general and cascade decompositions except that the state absent from the MV input literal has to have a different code from all the states present in that MV input literal in that same submachine.

An example of an encodeable selection of GPIs resulting in a two-way parallel decomposition is shown in Figure 6.15. As in the case of cascade and general decompositions, a lemma can be stated regarding the necessity and sufficiency of the encodeability constraints for parallel decompositions.

Lemma 6.6.2 *The constraints described in the above section are necessary and sufficient to ensure that a selection of GPIs satisfying them*

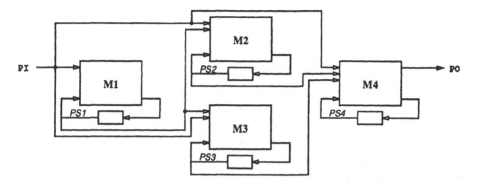

Figure 6.16: An arbitrary decomposition topology

results in a parallel decomposition that has the same functionality as the prototype machine.

Proof. The proof is similar to the proof for Lemma 6.5.1. □

6.6.3 Arbitrary Decompositions

An example of a topology that does not conform to any of the preceding topologies is shown in Figure 6.16. It is possible to target such topologies also by suitably modifying the covering constraints. The basic output constraint that any pair of states of the prototype machine should have distinct codes in the decomposed machine remains unchanged. The remaining constraints are dependent on the other submachines that a particular submachine receives present-state information from. The type of constraints imposed are briefly illustrated by means of the example topology in Figure 6.16. The constraints for submachine M_1 are the same as those for the head machine of a cascade chain. In the case of submachine M_2, if two states that have the same code in M_1, are given the same code in M_2 also, then all the next-state pairs of these two states should also be given the same codes in M_2. Similarly, if a state is absent from a MV input literal of M_2, it should have a different code from the states present in that MV literal in either M_1 or M_2. The justification for this constraint is similar to that for cascade and parallel decompositions. The constraints for M_3 are identical to those for M_2. The constraints for M_4 are similar.

In general, it is possible to target *any* desired topology by formulating the decomposition problem as a covering problem and suitably modifying the covering constraints.

6.6.4   Exactness of the Decomposition Procedure

Lemmas 6.6.1 and 6.6.2 for the cascade and parallel decomposition cases, respectively, are similar to Lemma 6.5.1 for the case of two-way general decomposition. Similar lemmas can also be stated regarding the encodeability constraints for arbitrary topologies. These lemmas in conjunction with Lemma 6.5.2 show that the decomposition obtained in this manner is actually the minimum cardinality solution.

6.7   Heuristic General Decomposition

6.7.1   Overview

The basic iterative strategy that has been used successfully for the two-level Boolean minimization problem appears promising for general decomposition also. The encodeability requirements for the selected GPIs are defined in the same manner as for the exact procedure (*cf.* Section 6.5.6). But, instead of enumerating all the GPIs, one begins with a set of GPIs corresponding to L_a and L_b (*cf.* Section 6.5.2), and an attempt is made to reduce their count, while maintaining the validity of the GPI covers. Operations similar to the *reduce* and *expand* operations of MINI [46] and ESPRESSO [13, 81] can be performed in an effort to minimize the cover cardinalities.

The three basic steps in the iterative loop are given below in the order in which they are carried out:

- Symbolic-reduce

- Symbolic-expand

- Minimize covers and remove input constraints

The cost function that is used for the iterative procedure is the same as that used for the exact algorithm, namely, the sum of the cardinalities of the minimized one-hot coded submachines. The steps of the iterative procedure are repeated until the cost of the solution, given by this cost function, remains unchanged after a pass through the loop.

In the symbolic-reduce and symbolic-expand steps, an attempt is made to modify the symbolic-output tags of the GPIs that currently make up the covers of **Submachine a** or **b** so that the possibility of obtaining a cover with a lower cardinality than the cardinality of the cover at the beginning of the current pass of the loop is increased. The symbolic-reduce and symbolic-expand steps are analogous to the reduce and expand steps, respectively, in iterative two-level Boolean minimization. The reduce and expand steps attempt to modify the cubes making up the Boolean cover so that the possibility of getting a cover after the *irredundant* step, with a lower cardinality than the cardinality of the cover at the beginning of the current pass of the loop, is increased. While the heuristic decomposition procedure *does not* check that the cost of the decomposition does not increase during a pass through the loop, the symbolic-reduce and symbolic-expand steps are carried out based on intelligent heuristics that do usually lead to a reduction in the cost. All the same, a copy of the best solution obtained up to the current point is maintained. The solution of the iterative procedure is the best solution obtained up to the point when the procedure enters a local minimum that it cannot climb out of. The steps of the basic iterative loop are explained below.

6.7.2 Minimization of Covers and Removal of Constraint Violations

In the minimization step, which follows symbolic-expand, a two-level minimization of the covers for both the submachines is carried out. This step incorporates *all* the cube-merging that becomes possible as a result of the symbolic-expand. The cover produced as a result of the minimization is termed the *over-minimized* cover because the minimization is carried without taking into account violations of the input constraints, and therefore may *not* be encodeable. The conditions giving rise to such a violation were elucidated in Section 6.5.8. When a violation of a MV input constraint is detected, the GPI with the constraint violation is split into two new cubes along the MV input part so that the MV input part of the first new cube is contained entirely within one of the two cliques (*cf.* Section 6.5.8) and the MV input part of the second new cube intersects only the remaining clique (does not intersect the other clique). This method of splitting the cube is not unique but is effective because at least one of the new cubes is devoid of any input constraint violations and the cardinality of the cover is only increased by one. The end result of this *unraveling* of the MV input literals of the GPIs is to make the over-minimized cover encodeable.

6.7.3 Symbolic-expand

The goal of the symbolic-expand procedure is to increase the size of the output tags of the GPIs in each cover till some form of primality is achieved. The cover is considered to be *prime* when no symbol can be added to the output tag of *any* GPI without violating the constraint that two states cannot be in the same output tag in the GPIs of both submachines (*cf.* Section 6.5.6). The atomic operation in the expansion procedure inserts two states in the same output tag of a GPI, checking while doing so that the output constraint is not violated (*cf.* Section 6.5.8). This operation is carried out alternately between the two submachines. The state pair is given a label corresponding to the submachine with which it is associated. Three non-trivial cases are possible when a state pair is selected for insertion:

- The state pair intersects no clique (*cf.* Section 6.5.8) with the same label in the encodeability graph and if it intersects a clique with a different label, the cardinality of that intersection equals one. If that is the case, a new clique is added to the graph with the same label as the state pair.

- The state pair intersects one clique in the graph with the same label and the cardinality of that intersection is one. In this case, the state pair is merged with that clique if the creation of the new clique (the label of the clique is unchanged) is such that no output constraint is violated, otherwise the state pair cannot be inserted.

- The state pair intersects two cliques with the same label (the cardinality of each intersection is one because no output constraint is violated up to the current point). In this case, the two cliques are merged into a single new clique with the same label if the new clique does not violate an output constraint. If an output constraint is violated upon merging, the state pair cannot be inserted.

The effectiveness of the expansion procedure depends a great deal on the order in which the state pairs are inserted into the same tag. Inserting a pair of states, say s_1 and s_2, in the same tag of a GPI in, say, **Submachine a** can affect the cardinality of the overall cover, after the cover has been minimized, in two ways:

1. Cubes that asserted the next-state s_1 and cubes that asserted the next state s_2 in **Submachine a** may merge, making the total number of cubes that assert the next states s_1 and s_2 smaller than before this expansion step. Thus, an expansion step can lead to a reduction in the cardinality of the cover (in **Submachine a or b**) in which the expansion is carried out.

2. The minimization step generates an over-minimized cover which is required to be made encodeable with respect to the input constraints. The insertion of a pair of states into the same output **tag** of a GPI in **Submachine a** generates *new* input constraints for *both* **Submachine a** and **b**. To make the cover encodeable, the MV input parts are unraveled (*cf.* Section 6.7.2), thereby invalidating some of the merging of cubes that occurred in the minimization step. Thus, the cardinality of the cover after unraveling could be greater than after the minimization step. It is possible for the increase in cardinality due to unraveling to be greater than the reduction in cardinality due to minimization *if* the choice of the state pair is inappropriate. In such a situation, the expansion step would, actually, lead to an increase in the overall cardinality rather than a reduction. Therefore, it is important to make the appropriate choices during expansion so that a reduction in cardinality may be achieved.

To maximize the possibility of a reduction in the cost of the cover, the state pairs are ordered at the beginning of every symbolic-expand using heuristics that reflect these two effects. The expansion step attempts to insert pairs with large weights first.

One ordering strategy tries to maximize the reduction in cardinality during the minimization step by assigning large weights to next-state pairs that have a large number of common present-states in their fanin and are asserted as a result of the application of a large number of common primary input combinations.

A second ordering strategy tries to minimize the number of new input constraints (*cf.* Section 6.5.6) that are generated in both **Submachine a** and **b** as a result of inserting a state pair into the same output tag, and to minimize the increase in cardinality due to unraveling. This is done by giving large weights to present-state pairs that assert the same next states and primary outputs. This heuristic is based on the fact that satisfying the input constraint due to a 1 in the positions corresponding to both the states, s_1 and s_2, and a 0 in the position corresponding to the state s_3 in the multiple-valued part of some GPI is identical to satisfying the output constraint for these three states if the two states, s_1 and s_2, occur together in the output tag of some GPI in either **Submachine a** or **b**. Because the output constraints are satisfied during the symbolic-expand step itself, no unraveling of the MV input part is required and the cardinality does not increase. Since two states asserting the same primary outputs and next states are likely to be in the same MV part in the minimized cover, an attempt is made to keep them in the same output tag.

6.7.4   Symbolic-reduce

The symbolic-reduce operation transforms the prime (*cf.* Section 6.7.3) cover into a non-prime cover. This operation is essential to the iterative process for moving out of the local minimum that it may have entered following the symbolic-expand and minimization steps. The basic operation used by symbolic-reduce in converting a prime cover to a non-prime cover is to remove a state from the symbolic output tags that it is contained in, when the tags contain more than one state in them, while maintaining functionality. The states selected for removal are those whose insertion in the output tags of GPIs during the symbolic-expand generated new input constraints. Because the input constraints are generated due to non-null intersections between cliques with different labels (*cf.* Section 6.5.8), it is ensured that after the symbolic-reduce, the intersection between *any* pair of cliques is null. Such a cover is said to be *maximally reduced.* This formulation of the symbolic-reduce operation is order independent.

6.8 Relationship to State Assignment

Early work in FSM decomposition and state assignment recognized the relationships between FSM decomposition and state assignment (*e.g.* [44]). Partition theory was used to both decompose and encode a sequential machine. However, the encoding step was simplistic in that the cost function targeted did not correspond to a minimized implementation.

As the interest in automated combinational logic optimization grew in the '70s and '80s, the need for better predictors in state assignment was seen. Multiple-valued minimization followed by input constraint satisfaction in [73], described in Chapter 3, represented a significant advance in deterministically predicting the complexity of the eventual implementation. However, at this stage the relationships between FSM decomposition and state encoding via input/output constraint satisfaction were obscured slightly, since the cost functions targeted by the procedures available for the two approaches were quite different. An effort to correlate the extraction of factors in a machine to guaranteed reduction in minimized product-term count over minimized one-hot coded implementations was made in [29]. Correlations were successfully made for certain classes of factors.

The generalized prime implicant (GPI) formulations for the state assignment and FSM decomposition problems bring out the strong relationship between FSM decomposition and state assignment. There are only slight differences in the definition of GPIs, and in the definition of encodeability. This is mainly because the goal of decomposition is to be able to implement a FSM as partitioned logic (partitioned by a cutset of latches), whereas no such restriction is applied during state assignment. The difference in the application of constrained covering to decomposition and state assignment lies in the step that checks for the encodeability of a selected set of GPIs. In fact, the restriction on the partitioning of logic between the submachines makes the encodeability check in FSM decomposition problem easier than that of state assignment.

To elaborate, consider the two-way decomposition problem. The goal of two-way decomposition is merely to find two partitions on the states where each partition could consist of a number of blocks. Decomposition does not carry out the complete encoding of the states, it merely "pre-processes" the states so that the subsequent state encoding applied on this pre-processed set of states will be guaranteed to realize the decomposition with the desired topology. As a result, the constraints involved in two-way decomposition are much simpler than the constraints that have to be checked for optimum state encoding, and the two-way decomposition problem is simpler than optimum state encoding. Note that the constraints become more complicated in three-way decomposition, and successively more complicated in four-way and five-way decomposition.

State encoding can be viewed as the problem of finding the optimal general decomposition of the prototype machine into as many submachines as there are state bits in the final state encoding. Based on this premise, one approach to making state encoding tractable for large problem sizes is to decompose the STG prior to the actual encoding. The decomposition makes the subsequent encoding steps much more tractable, and an efficient initial decomposition is one that does not compromise the optimality of the final result.

6.9   Experimental Results

The motivation for FSM decomposition was elucidated in some detail in Section 6.1. Based on the properties desired, the efficacy of a decomposition can be judged from the following criteria:

- The sum of the areas of the two-level (multilevel) implementation of the encoded submachines compared to that of the two-level (multilevel) implementation of the encoded prototype machine.

- The areas of the two-level (multilevel) implementation of each encoded submachine compared to that of the two-level (multilevel) implementation of the encoded prototype machine.

- The total number of inputs and outputs in each encoded submachine compared to the number in the encoded prototype machine.

Example	States	PI	PO	O-H Card	Enc. Bits	Enc. Card[1]	PLA Area[1]
bbara	10	4	2	34	4	24	528
bbsse	16	7	7	30	4	29	957
bbtas	5	2	2	16	3	8	120
beecount	7	3	4	12	3	10	190
dk27	7	1	2	10	3	9	117
dk512	15	1	3	21	4	20	340
ex4	14	6	9	21	4	15	465
fs1	67	8	1	583	7	119	4522
fstate	8	7	7	22	4	16	528
modulo12	12	1	1	24	4	12	180
s1a	20	8	0	92	5	73	2263
scf	121	27	54	151	7	137	17947
scud	8	7	6	86	3	62	1798
styr	30	9	10	111	5	94	4042
tav	4	4	4	12	2	10	180
tbk	32	6	3	173	5	57	1710

[1]Using NOVA for state assignment.

Table 6.1: Statistics of the encoded prototype machines

The heuristic procedure for optimal two-way general decomposition has been implemented in a program called HDECOM [4]. The input to HDECOM is a State Table description of the prototype machine. As output, HDECOM can produce either a fully encoded decomposed machine or a decomposed machine in which the present-state inputs to the submachines and their next-state outputs are symbolic. The heuristic algorithm in HDECOM has been tested on a number of examples obtained from the MCNC FSM benchmark suite [64] and industrial sources. The statistics of the chosen examples are shown in Table 6.1 (**O-H** stands for one-hot). The first step in the decomposition process is to obtain the STG representations of the submachines into which the prototype machine is decomposed. The second step involves the implementation of the individual submachines, *i.e.* an encoding of the states of the subma-

Example	States	Area[1] of Decomposed M/C[2]	Lumped M/C[1]	
			KISS	NOVA
bbara	4/3	360/180/**540**	650	**528**
bbsse	10/2	950/168/**1118**	1053	**957**
bbtas	3/2	91/60/**151**	195	**120**
beecount	4/2	126/105/**231**	242	**190**
dk27	4/2	55/30/**85**	117	**117**
dk512	8/2	238/104/**342**	414	**340**
ex4	4/4	128/325/**453**	589	**465**
fs1	16/5	3430/1254/**4684**	5510	**4522**
fstate	5/2	450/39/**489**	726	**528**
modulo12	6/2	117/44/**161**	225	**180**
s1a	8/3	1246/616/**1862**	2263	**2263**
scf	90/2	14832/3135/**17967**	18760	**17947**
scud	4/2	1050/720/**1770**	2698	**1798**
styr	16/2	3589/728/**4317**	4186	**4042**
tav	2/2	75/75/**150**	180	**180**
tbk	16/2	1275/425/**1700**	4410	**1710**

[1]PLA Area.
[2]Encoding for each submachine obtained using NOVA.
The first number in a column is the value for the
the first machine, the second for the second machine
and the third for the overall decomposed machine

Table 6.2: Results of the heuristic two-way decomposition algorithm

chines and a minimization of the resulting logic. In Table 6.2, the statistics of the final two-level implementations for the decomposed machines are shown. The encoding of the states of the submachines was obtained using NOVA [94]. Also shown in Table 6.2 are the areas of the logic-level implementations of the undecomposed prototype machines. The logic-level implementations of the prototype machines were obtained using two different state-encoding strategies: (1) KISS [73] and (2) NOVA [94]. NOVA is a state encoding program that produces competitive results for state assignment targeting two-level implementations. NOVA typically produces better results than KISS. It can be observed from Table 6.2 that the overall area of the decomposed machine is better than the area of the prototype machine implemented using KISS. It can also be observed that the implementation of the decomposed machine compares favorably in area to the prototype machine encoded using NOVA. Thus, the goal of decomposing FSMs and at the same time keeping a tab on the cost of the resulting logic-level implementation can be achieved, using the decomposition methods described. In the decompositions shown in Table 6.2, the individual submachines are generally much smaller in area than the prototype machine, illustrating that at least in the case of two-level implementations, a decomposition of the FSM translates directly into improved performance. The data corresponding to the performance characteristics of the multilevel implementations of the prototype and decomposed machines is shown in Table 6.3. The delay was computed using the unit-delay model in MIS-II [10], and is effectively the number of levels in the circuit. A performance improvement can be observed for most of the examples in the table.

The exact algorithm for decomposition was also implemented in [4] and the results are reported in Table 6.4. The same encoding strategies as used in the case of heuristic decomposition have been used to compare the implementations of the decomposed and prototype FSMs. The exact algorithm is not viable for large problem instances because the number of GPIs tends to be extremely large, resulting in excessive memory requirements and an intractable covering problem. However, it provides the formalism necessary for good heuristic approaches.

Because the submachines in a decomposition could have common inputs, extra routing area is required, over and above the PLA area. Typically, this extra area is small in comparison to the PLA areas and does not offset the area gain via decomposition. In addition, a submachine may actually be independent of some of the primary input and present-state lines, thus reducing the number of inputs required to be routed to that submachine. An example of such a case would be a cascade decomposition in which the operation of the head machine is completely independent of the present state of the tail machine.

Since the logic for a primary output or next-state line is entirely contained in a single submachine, it is apparent that the decomposition

Example	Decomposed Delay[1]	Lumped Delay[1]
bbara	4	6
bbsse	6	10
bbtas	3	4
beecount	4	6
dk27	3	4
dk512	4	7
ex4	3	6
fstate	4	9
modulo12	2	5
sla	5	7
scud	6	8
tav	2	4

[1]Using the unit-delay model in MIS-II.

Table 6.3: Performance improvement in multilevel logic via decomposition

Example	States	Decomposed[1] M/C Area[2]	Lumped M/C Area[1]	
			KISS	NOVA
contrived	4/4	40/66/106	120	108
shiftreg	4/2	36/10/46	72	48
lion9	5/2	112/5/117	136	136

[1]Two-Level Area.
[2]Encoding for each submachine obtained using NOVA.
The first number in a column is the value for the
the first machine, the second for the second machine
and the third for the overall decomposed machine

Table 6.4: Results of the exact decomposition algorithm

does not add multiple levels of logic between latch inputs and outputs. The effect of decomposition is therefore similar to partitioning the set of primary outputs and next-state lines in the overall FSM into two (or more in case of decomposition into multiple submachines) groups

Example	Literals in Decomposed M/C[1]	Literals in Lumped M/C[1]	
		NOVA	JEDI
bbara	37/26/63	59	54
bbsse	100/18/118	103	99
bbtas	15/10/25	22	22
beecount	21/13/34	32	33
dk27	17/6/23	25	22
dk512	41/25/66	53	50
ex4	31/26/57	57	57
fstate	39/20/59	62	-
modulo12	18/9/27	22	22
s1a	176/59/235	244	130
scud	101/97/198	195	-
tav	12/13/25	25	26

[1]After optimization using MIS-II.
The first number in a column is the value for the
the first machine, the second for the second machine
and the third for the overall decomposed machine

Table 6.5: Comparison of literal counts of multilevel-logic implementations

and implementing the logic driving each group of primary output and next-state lines separately. Such a partitioning of logic is termed **vertical partitioning**. In general, it is not necessary that a good vertical partition should exist for an arbitrary implementation of a FSM. [2] The decomposition procedure, on the other hand, can ensures that a good vertical partition does exist.

Reported in Table 6.5 are the multilevel logic literal counts for some examples for which a two-level implementation may not be efficient. Comparisons to the programs NOVA and JEDI [61] are included. It can be seen that the multilevel area of the decomposed machine is comparable

---

[2]Experimental results using vertical partitioning on FSMs encoded using various state assignment programs indicate that partitions that decrease overall machine area are not usually found.

to that of the best multilevel implementation of prototype machine for these examples, even though the decomposition strategy described is not geared specifically toward decomposition for optimizing multilevel area. The results imply that a good decomposition targeting two-level area can be a good decomposition for the multilevel case.

These results are significantly better, in general, than those obtained via factorization [29] due to a different problem formulation and the targeting of an improved cost function.

6.10   Conclusion

Various strategies for the heuristic and exact, multi-way general decomposition of finite state machines were described in this chapter. These algorithms are based on different cost functions which vary in how well they reflect the cost of a logic-level implementation. They also target different kinds of topologies of interaction between the submachines. The heuristic strategies presented that were based on the notion of generalized prime implicants have been implemented, and appear to perform reasonably well on benchmark examples.

Decomposition provides a method for simplifying the process of constraint generation for state encoding. The complexity of the optimal constraint generation problem for state encoding grows exponentially with the problem size. Decomposition by symbolic-output partitioning provides a way of simplifying the constraint generation problem. Rather than impose constraints on each bit of the code for every state (as would be done in state encoding), decomposition allows constraints to be imposed on groups of bits at a time. The actual encoding for each of these groups can subsequently be obtained in the second step. The heuristics described in this chapter for generating constraints to be imposed on groups of bits can be improved. It is possible that combining factorization (*cf.* Section 6.4) with symbolic-output partitioning could lead to superior results.

Chapter 7

Sequential Don't Cares

7.1 Introduction

Interacting finite state machines (FSMs) are common in chips being designed today. The advantages of a hierarchical, distributed-style specification and realization are many. While the terminal behavior of any set of interconnected sequential circuits can be modeled and/or realized by a lumped circuit, *i.e.* a composite machine, the former can be considerably more compact, as well as being easy to understand and manipulate.

The sequential logic synthesis algorithms described in the previous chapters are generally restricted to operate on lumped circuits. The decomposition methods typically begin from a State Transition Graph description of the required terminal behavior and produce several smaller State Transition Graphs corresponding to the decomposed submachines.

Reducing or flattening a given set of interacting machines into a single, composite machine could result in astronomical memory requirements and expend large amounts of CPU time. Unfortunately, given a set of interacting machines represented by State Transition Graphs, algorithms that encode the internal states of the machines, *taking into account their interactions*, do not exist to date. If indeed, the machines are encoded separately, disregarding their interconnectivity, a sub-optimal state assignment can result (and generally does).

One reason why the independent optimization of interacting

169

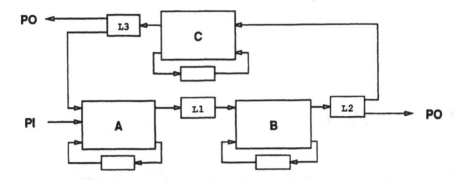

Figure 7.1: Interacting Finite State Machines

machines results in suboptimal designs is that don't care conditions that exist at the boundaries of any two interacting machines are ignored during the optimization. In this chapter, we will deal with the important issue of the specification and exploitation of don't cares in interconnected FSM descriptions.

Consider the interconnection of Figure 7.1. Certain binary combinations may never appear at the set of latches $L1$. This will correspond to an incompletely specified machine B. These don't cares can be exploited using standard incompletely-specified state minimization strategies (*e.g.* [76]). A more complicated form of don't care, referred to here as a *sequential don't care*, corresponds to an input sequence of vectors, say 1111, 1011, 1000 that does not appear at $L1$, though each of the separate vectors do appear. Sequential don't cares are more difficult to exploit. These don't cares are due to the limited controllability of B and can be used to optimize B. There are also other don't cares related to the limited observability of A.

In Section 7.2, we describe algorithms for state minimization of completely and incompletely-specified FSMs. In Section 7.3, we show the existence of input don't care sequences for a tail machine, driven by a head machine, in a cascade. We present a systematic method of state minimizing the machine under an input don't care sequence set that was originally published in [24]. Early work by Unger [93] and Kim and Newborn [51] on input don't care sequences is briefly reviewed. Input don't care sequences are the result of the limited controllability of a driven machine in a cascade. In Section 7.4, we show that output don't care sequences could result due to the limited observability of a driving machine in a cascade. We present a procedure [24] to exploit output don't cares. Exploiting either input or output don't cares can reduce the complexity of the driven and driving machines. A set of interacting FSMs can be iteratively optimized using these don't care sets.

The sequential don't cares described in Sections 7.3 and 7.4 can be utilized to reduce the number of states in a State Transition Graph (STG) description of a FSM. We will describe optimization procedures that utilize don't cares at the logic or encoded STG level. We focus on single FSMs in Section 7.5 and interacting FSMs in Section 7.6. The don't cares that can be utilized at the logic level correspond to a superset of the don't cares described in Sections 7.3 and 7.4.

There are many interesting relationships between the procedures described here and the non-scan sequential testability of a sequential circuit. The interested reader is referred to [3, 28, 27] for details. Our focus here is on simplifying the logic for area or delay minimization using these don't cares.

7.2 State Minimization

In this section, we will describe methods for the state minimization of completely and incompletely-specified FSMs. While the former problem is polynomial-time solvable, the latter has been shown to the NP-complete [35].

7.2.1 Generating the Implication Table

The first step in state minimization is to determine equivalences or compatibilities between each pair of states, by means of an implication table. This procedure is essentially the same for completely or incompletely-specified machines. We will illustrate this procedure with an example corresponding to the completely-specified STG of Figure 7.2.

We construct an implication table for the STG as shown in Figure 7.3. Vertically down the left side of the table all the states in the STG are listed except the first, and horizontally across the bottom all the states are listed except the last. Given a pair of states s_i, s_j we

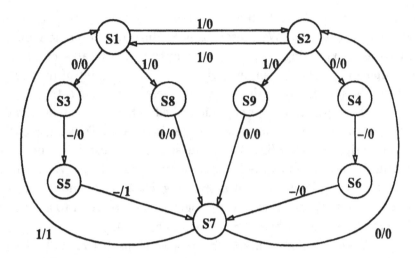

Figure 7.2: State Graph to be minimized

compare the next states and outputs corresponding to these states on each member of the input alphabet. We start by placing an **X** in each square corresponding to a pair of states which have different outputs and any input — these states cannot be equivalent or compatible. If the next state or the output from a particular edge is not specified for s_i, we assume that the next state or output is equal to that for s_j. In our completely-specified machine of Figure 7.2, state $s5$ is not equivalent to any state, and state $s7$ is not equivalent to any state.

We now enter in each square the pairs of states *implied* by the pair of states corresponding to that square. We start in the top square of the first column, corresponding to the pair of states $s1$, $s2$. From the STG of Figure 7.2, we see that this pair implies the pair $s3$, $s4$ which correspond to the fanouts on the 0-input from $s1$, $s2$, and the pair $s8$, $s9$ which correspond to the fanouts on the 1-input from $s1$, $s2$. We thus enter $3 - 4$ and $8 - 9$ in the square. We proceed in this manner till the table is complete. Note that we have a $\sqrt{}$ in the square $s8$, $s9$. This is because $s8$, $s9$ imply themselves on the 1-input and have identical fanout on the 0-input.

The next step is to make a "pass" through the implicant table crossing out any pairs that are ruled out as state equivalence pairs by the squares that have been marked with a **X**. In general, if there is an

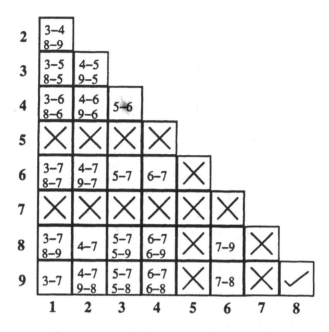

Figure 7.3: Initial implicant table

X in a square with coordinates $j - k$ then any square containing $j - k$ or $k - j$ as an implied state pair may be X'ed out. Since s_5 and s_7 cannot be equivalent to any other state, we find that we can cross out many squares — for instance, the squares $s1, s3$, $s1, s6$, $s1, s8$ and $s1, s9$ in the first column. The result after this pass is shown in Figure 7.4(a).

We make another pass through the table, since we have crossed out pairs in the previous pass, to check if we can cross out any more pairs. We find that we can cross out all the remaining squares in this pass, as shown in Figure 7.4(b). Thus, the only equivalences in this table correspond to $s8, s9$.

In general, we will keep making passes through the table until we do not cross out any squares in any given pass. When that happens, the procedure terminates. The squares that are not crossed out are equivalence pairs for a completely-specified STG, or compatible pairs for an incompletely-specified STG.

(a)

(b)

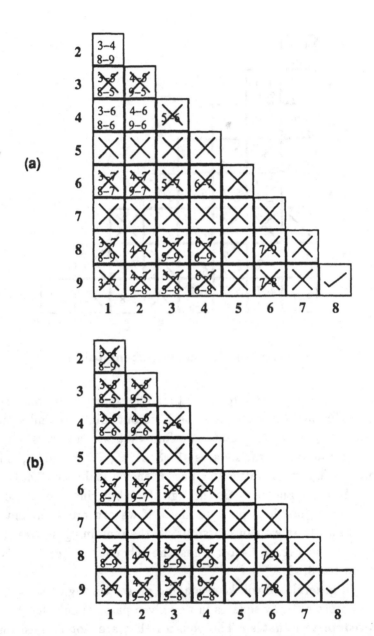

Figure 7.4: Implicant tables after first and second passes

7.2.2 Completely-Specified Machines

At the end of the procedure described in Section 7.2.1, in the case of completely-specified machines, we have a set of equivalence pairs.

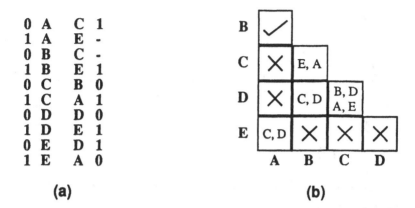

0	A	C	1
1	A	E	-
0	B	C	-
1	B	E	1
0	C	B	0
1	C	A	1
0	D	D	0
1	D	E	1
0	E	D	1
1	E	A	0

(a)

(b)

Figure 7.5: **Minimizing an incompletely-specified machine**

It has been shown that equivalence forms a class in completely-specified machines that if $s_i \equiv s_j$ and $s_j \equiv s_k$, then $s_i \equiv s_k$ [54]. This implies that the equivalent pairs can be merged into disjoint subsets of states. Each equivalence class corresponding to a disjoint subset of states can be replaced by a single state in the reduced machine.

Thus, a completely-specified machine can be reduced to minimum state form in polynomial time. A $n \log(n)$ algorithm to minimize the number of states in a sequential machine is due to Hopcroft [47].

7.2.3 Incompletely-Specified Machines

The problem with incompletely-specified machine state minimization is that compatibility is not a transitive relation. That is, if $s_i \cong s_j$ and $s_j \cong s_k$, then it is not necessary that $s_i \cong s_k$. A set of states is compatible if and only if every pair of states in the set is compatible.

This causes several complications after the compatible pairs of states have been derived using the procedure outlined. Consider the STT of Figure 7.5(a), taken from [54]. We have two edges with unspecified outputs. If we use the procedure described in Section 7.2.1 to derive an implicant table, we finally obtain the implicant table shown in Figure 7.5(b). Two points are worthy of note:

- As mentioned at the beginning of the section, compatibility is not a transitive relation. We have the compatible pair (A, B) and the pair B, C but states A and C are not compatible.

```
0 (A, B)   C        1
1 (A, B)   E        1
0 C        (A, B)   0
1 C        (A, B)   1                0 (A, E)    (B, C, D) 1
0 D        D        0                1 (A, E)    (A, E)    0
1 D        E        1                0 (B, C, D) (B, C, D) 0
0 E        D        1                1 (B, C, D) (A, E)    1
1 E        (A, B)   0
```

(a)   **(b)**

Figure 7.6: Two minimized realizations of the incompletely-specified machine

- Picking a particular compatible pair may require that another be picked. For example, picking A, E requires that C, D be picked as a compatible pair.

Let us pick the compatible pair A, B. We cannot pick the compatible pair A, E because B, E does not correspond to a compatible pair. Similarly, we cannot pick B, D because A, D is not a compatible pair, and we cannot pick B, C because A, C is not a compatible pair. Finally, we cannot pick C, D because it implies A, E (and B, E is not a compatible pair). Thus, we have minimized the five-state machine to a four-state machine, shown in Figure 7.6(a).

Instead of picking A, B as the first compatible pair to be merged, let us pick A, E first. This means we *have* to pick C, D as a compatible pair, which in turn means that we have to pick B, D, and A, E which is already picked. Picking B, D implies that we have to pick the implied pair C, D (which is already picked) and B, C to complete the B, C, D triangle. Therefore, we have the state compatibility sets A, E and B, C, D, implying that we can reduce the machine of Figure 7.5(a) to a two-state machine as shown in Figure 7.6(b).

The above example demonstrates the non-uniqueness of the minimal machine in the case of incompletely-specified machines. As mentioned earlier, selecting state compatibility pairs or sets to result in a minimum-state machine is an NP-complete problem. Therefore, one has to resort to heuristics in an attempt to obtain minimal solutions within reasonable amounts of time for large machines.

A set of compatible states is said to be a **maximal compatible set** if it is not covered by any other compatible class. Similarly, a set of incompatible states is said to be a **maximal incompatible set** if it is not covered by any other incompatible class. Paull and Unger give methods to generate maximal compatibles in [76].

When the number of compatibility pairs or maximal compatible sets is not very large, a brute-force method of enumerating all choices in compatible set selection is reasonably efficient. Further, maximal compatible and incompatible sets serve as good guides to the heuristic state minimization of incompletely-specified machines. The use of maximal compatible and incompatible sets in a relatively simple heuristic selection process suffices to obtain minimal realizations, in most cases, on real-life, incompletely-specified machines.

7.2.4 Dealing with Exponentially-Sized Input Alphabets

One problem faced in the synthesis of VLSI FSMs that is generally ignored in the early work on FSM optimization is that the input alphabet of the sequential circuits being optimized is of size 2^{N_i} where N_i is the number of inputs to the sequential machine. N_i can be quite large for single FSM controllers. The procedure of implicant table construction as described in Section 7.2.1, requires the explicit comparison of the next state/output of each fanout edge from each state to the corresponding edge on the same primary input minterm from every other state. This implies 2^{N_i} comparisons for a completely-specified machine.

Typically, even if the machine has a large number of inputs, the number of specified edges is small because the edge inputs have been specified as cubes, rather than minterms. However, we may have a situation where state s_1 in the machine with 3 inputs has fanout edges on $1-1$, $0-1$, and $--0$, but state s_2 has fanout edges on 111, 110, 10- and $0--$. In this case, a one-to-one correspondence between the edges of s_1 and s_2 is not possible.

One solution is to create an input alphabet for the machine corresponding to a disjoint set of primary input cubes $c_1, \cdots, c_P$, where $i \neq j \Rightarrow c_i \cap c_j = \phi$. This can be done by using algorithms that make a two-level cover disjoint (*e.g.* [13]). Then the fanouts of every state in the machine can be expressed under the input alphabet corresponding to $c_1, \cdots, c_P$. In many cases, P can be substantially smaller than 2^{N_i}, but for large N_i, P can be quite large.

An alternate, more efficient method is to leave the specification of the machine untouched, but instead perform intersections on the edge input cubes. For instance, given two states s_i and s_j, we intersect the inputs corresponding to each pair of fanout edges from s_i and s_j, *i.e.* perform the operation $e_k.Input \cap e_l.Input$. If $e_k.Input \cap e_l.Input \neq \phi$, then we know that there exists an edge pair from s_i, s_j on the same input minterm within e_k, e_l, and we inspect $e_k.Next$, $e_l.Next$ and $e_k.Output$, $e_l.Output$ during implicant table generation. In our example above, we will compare the next state/output of s_1 on $1 - 1$ against the next state/output of s_2 on 111 and 10$-$. The number of comparisons for a given pair of states s_i, s_j is thus restricted to the number of fanout edges from s_i times the number of fanout edges from s_j.

A procedure that implicitly generates equivalence classes in a sequential machine using Binary Decision Diagram-based image computation methods [20] was described in [63]. [1] This procedure has been used to generate equivalence classes for machines with over 100 latches. However, this method is currently restricted to operate on completely-specified machines described at the logic level.

7.3 Input Don't Care Sequences

In Figure 1, we have a machine A driving another machine B via a set of latches $L1$ (We neglect C for the moment). For the purposes of the discussion here, we assume that all the latches in $L1$ are not directly observable. In practice, a subset of the latches may be directly observable. However, the don't care exploitation techniques described here are easily modified to the general case.

7.3.1 Input Don't Care Vectors

We assume that a State Transition Graph description exists for both machines A and B. Let the number of intermediate/pipeline latches in $L1$ be N. A may or may not assert all 2^N possible output

---

[1]Image computation methods can efficiently compute the set of fanout states for a given set of states. These methods have been used in sequential logic testing and verification and are described in a companion book [38].

combinations. If a certain binary combination, c_1 never appears at $L1$, then B will be incompletely-specified – the transition edges correspond-ing to an input of c_1 need not be specified, whatever state B is in (We don't care what happens when B receives the input c_1). The more gen-eral case is when a certain combination c_2 never appears at $L1$, when B is in some set of states $S_B \in Q_B$ (Q_B is the set of all states B can be in). It does appear when B is in states other than S_B. In this case, the states in S_B will have c_2 unspecified (If an edge on c_2 exists in S_B, it can be removed). This type of don't care can be easily exploited via the use of state minimization algorithms that handle incompletely-specified machines that were described in Section 7.2.

7.3.2 Input Don't Care Sequences to Minimize States

A more complicated sequential don't care is associated with vector *sequences* that never appear at $L1$, though all 2^N separate vectors appear. A does not produce all possible output sequences. This type of don't care does *not* have a straightforward interpretation. Edges in the State Transition Graph of B cannot be removed or left unspecified. In Figure 7.7, State Transition Graphs corresponding to a machine A (left) driving another machine B (right) are shown. The two outputs of A are (directly) the inputs to B. The starting states of the machines A and B are $q1$ and $s1$, respectively.

The machine B is state minimal in isolation. Each transition edge in B is irredundant, *i.e.* B makes every transition with appro-priate input sequences. However, A does not assert all possible output sequences. A don't care input sequence, namely (11, 11), is shown be-low the graph of B. Such a don't care sequence implies that certain *sequences of transitions* will not be made by B.

A don't care input sequence is assumed to have a length greater than 1. A sequence of vectors VS_1 is said to **contain** another sequence VS_2 (written as $VS_1 \supseteq VS_2$), if VS_2 appears in VS_1. Given a don't care sequence DC, all sequences SE such that $SE \supseteq DC$ are also don't care sequences.

Definition 7.3.1 : *An atomic don't care sequence is a sequence that does not contain any other don't care sequence.*

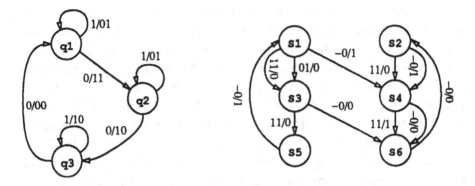

The sequence (11, 11) is a don't care.

Figure 7.7: Input don't care sequences

Thus, any subsequence of an atomic don't care sequence is a care sequence. In the sequel, we consider only atomic don't care sequences.

7.3.3   Exploiting Input Don't Care Sequences

Given a set of sequences that a driving machine never asserts, our problem lies in exploiting this form of don't care, so as to optimize B. In the general case, we will have a set of don't care sequences. Using the notion of delayed-input realizations, treated in [45] (p. 149), these don't cares can be converted to combinational don't cares, and then exploited using classical state minimization algorithms. Here, we are concerned with exploiting these don't cares to optimize the *given* sequential circuit. We can state the following lemma [24].

Lemma 7.3.1 : *Given a machine B and a set of don't care sequences $DSET = \{DC_1, DC_2, \cdots DC_{N_A}\}$, if the set of all differentiating sequences of two states $q_1, q_2 \in Q^2{}_B$ is $ISET = \{I_1, I_2, \cdots I_{N_D}\}$, such that for each k, $I_k \supseteq DC_l$ for some l, then q_1 and q_2 are compatible in B under $DSET$.*

Proof. *ISET* represents *all* the differentiating sequences of states q_1 and q_2. Since no $DC_j \in DSET$ can ever occur, it means the no $I_i \in ISET$ can ever occur, *i.e.* no differentiating sequence can be applied.

Therefore, q_1 and q_2 in B are compatible under $DSET$. □

In the example of Figure 7.7, states $s1$ and $s2$ have a single differentiating sequence (11, 11) and satisfy the conditions of Lemma 7.3.1 under the don't care input sequence (11, 11); they can therefore be merged into a single state. Note that a standard state minimization strategy will *not* alter machine B.

An approach to exploit don't cares based on Lemma 7.3.1 would entail producing *all* differentiating sequences for *every* pair of states in B and checking for the containment condition. Pairs satisfying the condition can be merged. This is potentially very time consuming; a pair of states may have many differentiating sequences and we have to find them for every possible pair. After finding these differentiating sequences and checking for the containment condition of Lemma 7.3.1, we have to solve a covering problem to minimize the number of states in the machine.

A more efficient approach is now outlined. In this approach, given a set of don't care sequences, B is transformed into a new machine B' which has a greater number of states, but is more incompletely-specified than B. B' is state minimized to obtain B'' ($||Q_{B''}|| \leq ||Q_B||$). The pseudo-code below illustrates the procedure.

In the above procedure, paths consisting of sequences of transition edges are enumerated from the initial state of B'. Each path P is checked to see if the input combinations corresponding to the path ($P.Input$) contain the don't care sequence DC_j. If not, depth-first enumeration continues. Assume that the DC_j has length $L = 2$ and $K = 2$. We will split the last state s_2 into two states s_2' and s_2'', initially duplicating fanout. The state s_2' that receives the last but one input, namely e_1 will have $e_2.Input$ unspecified. When $L > 2$, we can have $i = p < K$ in the *for* loop above. In this case, the fanout of s_p is duplicated for the states s_p' and s_p'' – the edge e_p is also duplicated. Hence, at the next iteration, one of the e_p fans into s_{p+1}' and the other e_p (as well as the remaining fanout edges from s_p' and s_p'') into s_{p+1}''.

The procedure is effectively producing a machine where the don't care sequences are *not* specified, but otherwise has the same functionality as the original machine. This means that if any two states in B satisfy the conditions of Lemma 7.3.1, these two states will not pos-

exploit-input-dc(B, DC):
{
    $B' = B$;
    foreach (don't care sequence DC_j) {
        $L = \|DC_j\|$;
        foreach (depth-first path $P = e_1, \ .. \ e_K \ \in B'$) {
            if ($P.Input \supseteq DC_j$) {
                for($i = K - L + 2; \ i \le K; \ i = i + 1$) {
                    $s_i = e_i.Present$;
                    make states s_i' and s_i'' ;
                    $fanin(s_i') = \{e_{i-1}\}$;
                    $fanin(s_i'') = fanin(s_i) - \{e_{i-1}\}$;
                    if ($fanin(s_i'') = \phi$) delete s_i'' ;
                    if ($i < K$)
                        $fanout(s_i') = fanout(s_i'') = fanout(s_i)$;
                    else {
                      $fanout(s_i') = fanout(s_i) - \{e_i\}$;
                      $fanout(s_i'') = fanout(s_i)$;
                    }
                    delete s_i ;
                }
            }
        }
    }
    $B'' =$ **state-minimize** (B');
}

Figure 7.8: Procedure to exploit input don't care sequences

sess a differentiating sequence in B' and will thus be *compatible* during standard, incompletely-specified state minimization. A smaller machine B'' will be obtained after state minimization.

    An illustrative example is given in Figures 7.7 and 7.9. The machine B and the don't care sequence of Figure 7.7 produce an expanded machine, shown in Figure 7.9(a). State minimizing this machine

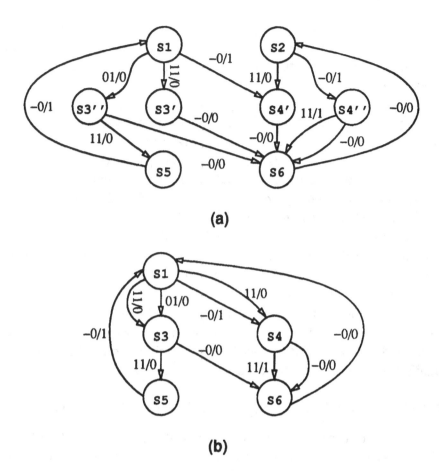

(a)

(b)

Figure 7.9: Example of exploiting input don't care sequences

produces the result of Figure 7.9(b), which has one less state than the
original machine of Figure 7.9. States $s3$ and $s4$ are split during expansion, however, states $s3'$, $s4'$ and $s4''$ merge and so do $s1$ and $s2$ during
state minimization.

Given a cascade, we need to generate the set of sequences that
the driving machine in a cascade $A \rightarrow B$ never asserts, so as to optimize
the driven machine B as in Figure 7.9. This is done by generating
don't care sequences of increasing length, beginning from a length of 2.
Starting from each possible state in A, all possible 2-vector sequences
are found. Single vectors that *don't* occur are added to this set and the
set is "complemented" to find the atomic 2-vector sequences that don't

occur. Next, all sequences of length 3 that A asserts are found. The single-vector and 2-vector don't care sequences are added to this set and the union is complemented to find the atomic don't care sequences of length 3 and so on.

For instance, consider a driving circuit where the single vectors 00, 01 and 10 appear. Therefore, the input sequences of length one that appear are $I_1 = \{00, 01, 10\}$. Assume that the input vector sequences of length 2 that occur are $I_2 = \{(00, 01), (00, 00), (00, 10), (01, 01), (01, 10), (10, 10)\}$. The atomic don't care input sequence set of length one is $\{11\}$, obtained by complementing I_1. We add $(11, --)$ and $(--, 11)$ to I_2 and complement the result to obtain the atomic don't care sequences of length 2, *i.e.* $\{(10, 01), (10, 00), (01, 00)\}$. If we do not add the single-vector don't care set to I_2, we will obtain a don't care input sequence set of length two that is not atomic, *i.e.* it will contain don't care input sequences of length one.

Expanding a machine under don't care input sequences of long length can be very time consuming. In the experiments described in [24], the don't care input sequences were restricted to length three, and significant optimization was obtained on a set of examples.

7.3.4 Early Work on Input Don't Care Sequences

In Unger's book [93], the notion of a pair-restricted state table is presented. A pair-restricted state table is one in which all the don't care entries can be accounted for by the existence of one or more disallowed input sequences of length two. Unger showed that the state minimization problem for a pair-restricted flow table is *easier* than the minimization of arbitrary, incompletely-specified sequential machines (p. 48-50). An example of minimizing a machine under a don't care sequence of length three is given.

A procedure tailored towards minimizing the number of states in the tail machine of a cascade of two machines is given by Kim and Newborn in [51]. The procedure uses algorithms derived from those described by Hennie [45] for the cascade decomposition of sequential machines. Essentially, given a cascade $M_1 \rightarrow M_2$, the procedure allows for the minimization of M_2 under the *complete* input sequence don't care set corresponding to M_1. The procedure of Kim and Newborn can be viewed as a two-step process where a product machine corresponding to the cascade $M_1 \rightarrow M_2$ is constructed, and M_1, the independent component is extracted out of the product guaranteeing a minimal-state M_2'. Note that the *useful* don't care sequence set will be bounded by the maximum length of an acyclic path in the product machine. Construction of the product machine requires knowledge of the starting states of the two machines.

Typically, one is concerned with the minimization of machines that are incompletely-specified due to don't care sequences of arbitrary, but finite length. Given a machine M_2 with an associated don't care input sequence set, the expansion step of the previous section is similar to the expansion illustrated by means of an example in Unger's book [93] (p. 51). The algorithm described here is less powerful and Kim and Newborn's procedure, because the Kim and Newborn procedure can take into account sequences that do not occur when the product machine is in a given state. However, the algorithm described here can be extended so as to not require knowledge of the starting state (as the procedure of [93]). The algorithm can be used to optimize for arbitrary sets of finite-length don't care sequences that are specified for *particular* states in M_2, and not for other states. This proves useful if one wishes to consider don't care sequences of small lengths in order to improve efficiency.

7.4   Output Don't Cares to Minimize States

The sequential don't cares discussed thus far are a product of the constrained controllability of the driven machine B in a cascade $A \rightarrow B$. There is another type of don't care due to the constrained observability of the driving machine A.

We focus on the individually state minimized State Transition Tables (STTs) of the cascade $A \rightarrow B$ of Figure 7.10. The intermediate inputs/outputs are symbolic. [2] Given that A feeds B, it is quite possible that for some transition edge $e_1 \in A$, it does not matter if the output asserted by this particular transition edge is, say, $INTi$ or $INTj$. In fact, in Figure 7.10, the 3rd transition edge of A can be either $INT1$ or $INT2$, *without changing the terminal behavior of* $A \rightarrow B$ (We assume

---

[2]For instance, the input alphabet for machine A is ($i1$, $i2$) and the output alphabet is ($INT1$, $INT2$).

I1	sa1	sa2	INT1		INT1	qb1	qb2	out1
I2	sa1	sa3	INT2		INT2	qb1	qb3	out2
I1	sa2	sa1	INT2		INT1	qb2	qb2	out3
I2	sa2	sa3	INT1		INT2	qb2	qb2	out3
I1	sa3	sa1	INT1		INT1	qb3	qb2	out4
I2	sa3	sa2	INT1		INT2	qb3	qb3	out1

A ——————————————————————▶ B

Figure 7.10: Example of output expansion

that there are no latches between A and B, the starting state of A is *sa1* and the starting state of B is *qb1*). This is a don't care condition on A's outputs. It is quite possible that making use of these don't cares can reduce the number of states in A. In fact, if one replaced the output of the 3rd edge in A (Figure 7.10) by $INT1$ instead of $INT2$, we would obtain one less state after state minimization (*sa2* becomes compatible to *sa3*).

7.4.1 Exploiting Output Don't Cares

Given a cascade $A \rightarrow B$, we give below a systematic procedure to detect this type of don't care, *i.e.*, to expand the output of each transition edge of A to the set of all possible values that it can take while maintaining the terminal behavior of $A \rightarrow B$. Standard state minimization procedures can exploit don't care outputs, represented as cubes. However, state minimization procedures have to be modified in order to exploit transition edge outputs represented as arbitrary Boolean expressions (multiple cubes) or arbitrary sets of symbolic output values in the output alphabet.

A transition edge e_1 in A is picked. The set of states that B can be in when A makes this transition is found. This is done by effectively traversing the product machine corresponding to $A \rightarrow B$, without explicitly constructing it. Beginning from the starting states of A and B all possible paths to $e_1 \in A$ are traversed, while simultaneously traversing paths in B. In the worst-case we will traverse the entire product machine. However, the memory requirements of this procedure are small. Given this set of states, the largest set of output combinations/symbols

output-expansion(A, B):
{
    foreach (edge $e_1 \in A$) {
        $OUT(e_1) = universe$;
        foreach (state $q_1 \in Q_B$) {
            if (B can be in q_1 when A makes transition e_1) {
                find largest set of output combinations c_1 |
                $c_1 \supseteq e_1.Output$ && $\lambda(c_1, q_1)$, $\delta(c_1, q_1)$ are unique ;
                $OUT(e_1) = OUT(e_1) \cap c_1$;
            }
        }
        $e_1.Output = OUT(e_1)$;
    }
}

Figure 7.11: Procedure to expand outputs

that covers the output of the edge and produces a unique next state and a unique output when B is in *any* one of the possible states is found (corresponds to $OUT(e_1)$). The output of e_1 is expanded to the set. The process is repeated for all edges in A.

The state minimization procedure described in Section 7.2 can be used for incompletely-specified finite state machines. However, after output expansion, we may have a multiple-output FSM in which a transition edge has an output that can belong to a subset of symbolic or binary values, rather than the universe of possible values (as in the incompletely-specified case).

In the state minimization procedure described in Section 7.2, two states are deemed to be compatible if the output combinations that can be asserted by each pair of corresponding fanout edges of the two states intersect. One can envision a situation where the possible output combinations of the fanout edges of q_1, $q_2 \in Q^2{}_M$ intersect leading to a compatibility relation $q_1 \cong q_2$, with similar compatibility relations $q_2 \cong q_3$ and $q_1 \cong q_3$. However, the three-way intersection between the possible output combinations of the fanout edges of q_1, q_2, and q_3 may be a null intersection, implying that q_1, q_2, and q_3 cannot be merged into a single state, even though all the required pairwise compatibility relations exist. In the binary-valued output case, if the possible output combinations can be represented as a single cube, then such a situation will not occur, since the three-way intersection of a set of three cubes has to be non-empty if the pairwise intersections are non-empty. But, in the case of multiple cubes or Boolean expressions specifying the output combinations for fanout edges, an additional check has to be performed during state minimization during the selection of the compatibility pairs to see if three or more sets of states can, in fact, be merged, preserving functionality.

When we have a set of interconnected machines as in Figure 7.1, the don't cares corresponding to each cascade can be *iteratively* used. For instance, in Figure 7.1, A drives B. The outputs of A's edges can be expanded first. A's output don't care sequences can be used to optimize B. Next, one can focus on $B \rightarrow C$. Output expansion can be performed on B and so on.

7.5 Single Machine Optimization at the Logic Level

7.5.1 Introduction

In this section, we begin with showing how invalid or unreachable states in a FSM can be explicitly specified as don't care conditions for the next state and output logic (Section 7.5.2). As described in Section 3.2, multiple-valued minimization during symbolic input or state encoding implicitly utilizes the unused code don't care set.

We will show in Section 7.5.3 how equivalent states in a STG can be specified as don't cares for the next state logic (originally described in [27]). This method alleviates the problem faced in state minimization that reduced machines don't necessarily result in optimal logic implementations. State splitting, *i.e.* having equivalent states in a given STG may be required for an optimal implementation [43].

```
                              0 00   00 1
                              1 00   01 0
     0 00   00 1              0 01   10 1              1 0-   01 0
     1 00   01 0              0 10   10 0              0 1-   10 0
     0 01   10 1              1 10   00 1              0 0-   00 1
     0 10   10 0              - 11   -- -              1 1-   00 1
     1 10   00 1              1 01   -- -              1 -1   10 0
        (a)                      (b)                      (c)
```

Figure 7.12: Invalid state and unspecified edge don't cares

7.5.2   Invalid State and Unspecified Edge Don't Cares

The utilization of invalid state don't cares is relatively straight-forward. Consider the State Transition Table (STT) of Figure 7.12(a). There are 3 states in the STT, which are reachable from the starting state 00. The table is a minimum product-term realization, assuming an empty don't care set. However, we note that the state 11 (not shown in the table) is not reachable from the starting state. The edges from 11 are thus not specified. Further, the edge on the 1-input from state 01 is not specified in the table.

We can add don't cares to the STT description before minimization as shown in Figure 7.12(b). Don't cares can be specified for both the next state logic and the output logic. Two-level minimization under the don't care set results in the minimized table of Figure 7.12(c).

7.5.3   Equivalent State Don't Cares

Equivalent/compatible states and state minimization were the subjects of Section 7.2. State minimization is an operation carried out at the symbolic level and suffers from the same drawbacks as state encoding — predicting the effects of merging states at the logic level is difficult. It has been known for a long time [43] that the optimal realization of a particular FSM may correspond to a non-minimal STG, *i.e.* a STG with pairs of states that are equivalent. In this section, we describe means of utilizing equivalent state don't cares to *directly* simplify the logic-level implementation of an encoded FSM. Using this method finesses the prediction problem in state minimization, to a certain extent.

We have a STT in Figure 7.13(a), whose encoded STT with

0	000	001	0
1	000	110	1
0	001	000	0
1	001	100	1
0	010	110	1
1	010	000	0
0	100	010	1
1	100	110	0
0	110	010	1
1	110	000	0
–	011	–––	–
–	101	–––	–
–	111	–––	–

0	S1	S2	0
1	S1	S5	1
0	S2	S1	0
1	S2	S4	1
0	S3	S5	1
1	S3	S1	0
0	S4	S3	1
1	S4	S5	0
0	S5	S3	1
1	S5	S1	0

1	–00	110 0
1	001	100 1
0	01–	110 1
0	000	001 0
0	1––	010 1

(a)   (b)   (c)

Figure 7.13: A State Graph with equivalent states

the invalid state codes 011, 101 and 111 specified as don't cares. is
shown in Figure 7.13(b). Note that the symbolic states $s3$ and $s5$ were
equivalent in the original STT and therefore the states 010 and 110
are equivalent. This can be considered a state-split realization. The
minimized implementation in Figure 7.13(c) has 5 product terms, and a
total of 25 literals (the 1's and 0's in the input plane and the 1's in the
output plane).

Let us state minimize the machine of Figure 7.13(a) to the machine of Figure 7.14(a). We have merged the states $s3$ and $s5$ into a
single state $s3$. We code the states in the machine into the encoded
STT, essentially with the same encoding as chosen in Figure 7.13(b),
(cf. Figure 7.14(b)). We have a larger invalid state don't care set, and a
smaller four-state machine, but the encoded and minimized implementation for this machine (cf. Figure 7.14(c)) has 6 product terms! Thus,
state minimizing a machine does not always result in smaller implementations.

We can exploit equivalences at the logic level by specifying a
next state don't care set or a fanin don't care set for the edges whose next
or fanin states are 010 or 110. The encoded STT of Figure 7.13(b) is
shown in the STT of Figure 7.15(a), except that the next state columns
which were 010 or 110 are now specified as –10. Minimizing this speci-

```
                          0 000   001 0
                          1 000   010 1
                          0 001   000 0
                          1 001   100 1
    0 S1  S2  0           0 010   010 1
    1 S1  S3  1           1 010   000 0
    0 S2  S1  0           0 100   010 1
    1 S2  S4  1           1 100   010 0          0 000   001 0
    0 S3  S3  1           - 011   ---  -         1 000   010 1
    1 S3  S1  0           - 101   ---  -         1 --1   100 1
    0 S4  S3  1           - 110   ---  -         0 1--   000 1
    1 S4  S3  0           - 111   ---  -         - 1--   010 0
                                                 0 -1-   010 1
         (a)                   (b)                   (c)
```

Figure 7.14: State minimization results in a larger implementation

```
                   0 000   001 0
                   1 000   -10 1
                   0 001   000 0
                   1 001   100 1
                   0 010   -10 1
                   1 010   000 0
                   0 100   -10 1
                   1 100   -10 0
                   0 110   -10 1         0 000   001 0
                   1 110   000 0         1 -00   010 0
                   - 001   ---  -        1 00-   100 1
                   - 101   ---  -        0 -1-   010 1
                   - 111   ---  -        0 1--   010 1
                        (a)                  (b)
```

Figure 7.15: An example of utilizing a next state don't care set

fication results in a minimized implementation (cf. Figure 7.15(b)) with 5 product terms and 22 literals – the best of the three different implementations.

Thus, specifying fanin don't care sets for equivalent states, and allowing the combinational logic optimization step to choose an appropriate state minimization represents an elegant means of finessing the problem of predicting logic complexity during state minimization (taking into account state splitting).

7.5.4   Boolean Relations Due To Equivalent States

Consider a scenario wherein two equivalent states in an encoded machine have been assigned the codes 100 and 111. In this case, we do not a simple don't care condition for the fanin edges to these states. 100 and 111 cannot be merged into a single cube. We have to specify this complex degree of freedom as a Boolean relation [16]. Minimization methods for Boolean relations have been proposed (*e.g.* [36, 88]) and are briefly described in the next section.

7.5.5   Minimization With Don't Cares and Boolean Relations

Optimizing a two-level circuit under a don't care set is a very well-understood problem. Programs like MINI [46] and ESPRESSO [81] can be used to obtain minimal or minimum covers under a specified don't care set.

Minimizing a Boolean relation, on the other hand, is much more difficult. An exact method for the minimization of Boolean relations was presented in [88]. The exact method uses the notion of candidate-primes that is similar to the generalized prime implicants presented in Chapters 4 through 6. Unlike exact two-level minimization, exact minimization of Boolean relations is only feasible for relatively small examples. A heuristic minimizer for Boolean relations viable for larger circuits was presented in [36]. Such a minimizer can be used to Boolean relations resulting from equivalent states.

The problem of multilevel minimization under external don't care sets has been a subject of recent investigation. Methods based on the iterative two-level logic minimization of each node in the given Boolean network were proposed in [7, 84]. More efficient methods that use Binary Decision Diagrams as a base representation were presented in [86].

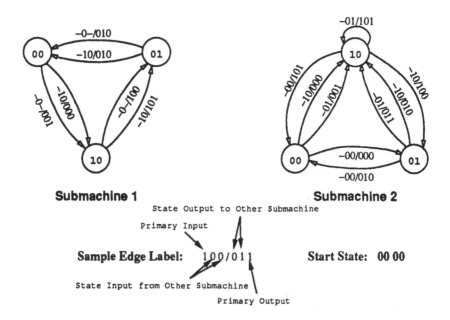

Figure 7.16: The State Graphs of two interacting machines

7.6 Interconnected Machine Optimization at the Logic Level

In this section, we present sequential don't care conditions and optimization procedures specifically suited to a class of interacting sequential circuits commonly found in practice. Two examples of such topologies are shown in Figure 6.2. The sequential circuit in Figure 6.2(a) consists of four interacting FSMs, while the circuit in Figure 6.2(b) consists of two such FSMs. The salient feature of this class of circuits is that communication between the FSMs is solely via their present states. This is the only requirement imposed. Usually, an interacting sequential circuit that is to be optimized but does not satisfy this property can be made to satisfy it by an appropriate repartitioning of the combinational logic. Cascade and parallel interactions between submachines are special cases of the interaction shown in Figure 6.2(b).

Don't care conditions are illustrated by means of the example of Figure 6.2(b). Shown in Figure 7.16 are the STGs of the two mutually-

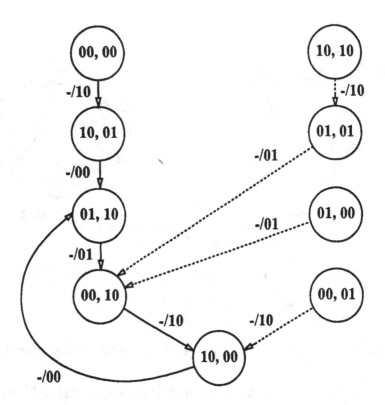

Figure 7.17: The State Graphs of the overall (product) machine

interacting FSMs of Figure 6.2(b). [3] The start state of the overall interacting circuit is 00 00, which corresponds to a start state of 00 for each individual FSM.

7.6.1 Unconditional Compatibility

Let us focus on states 00 and 01 in Submachine 2 of Figure 7.16. If this submachine is viewed in isolation, treating its outputs that feed Submachine 1 as primary outputs, then 00 and 01 are not compatible states. However, looking carefully at the edges of Submachine 1, we note that the state inputs 00 and 01 from Submachine 2 always produce the same response from Submachine 1. This implies that we

---

[3]Had the topology been a cascade, $M_1 \rightarrow M_2$, the state input to M_1 from M_2 would be all don't cares.

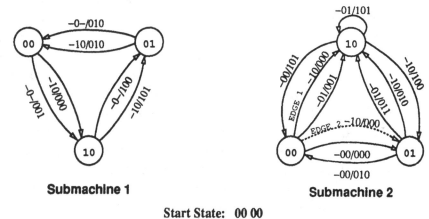

Submachine 1

Submachine 2

Start State: 00 00

Figure 7.18: Conditionally compatible states

can minimize the states in Submachine 2 from three to two, if we detect that states 00 and 01 are compatible *given that Submachine 2 interacts with Submachine 1*.

We call the type of compatibility above, **unconditional compatibility**. Unconditionally compatible states in a submachine can be merged into the same state. Looking at the STG of the overall product machine (for Submachine 1 and 2) in Figure 7.17, we see that for all $x \in \{0, 1\}$ and $y \in \{0, 1\}$ the states $xy, 01$ and $xy, 00$ are equivalent states in the completely-specified product machine. Note that some of these states are unreachable or invalid states.

7.6.2 Conditional Compatibility

There is a subtler form of don't care corresponding to the **conditional compatibility** of states in a submachine. In Figure 7.18, consider the situation where the edge from state 00 in Submachine 2, namely *EDGE1* that fans out to state 10 is moved to state 01 instead (shown by the dotted edge *EDGE2*). Note that this does not change the functionality of the overall circuit. However, 10 and 01 in Submachine 2 are not unconditionally compatible states. Moving other fanin edges of 10 to 01 *will* change the functionality of the overall circuit.

To understand conditional compatibility we go back to the STG of the product machine shown in Figure 7.17. Note that the state $01, 01$ in the product machine is equivalent to the state $01, 10$. However, the state $00, 01$ is not equivalent to the state $00, 10$, implying that we cannot merge 01 and 10 in Submachine 2. Thus, conditional compatibility is defined with respect to a subset of fanin edges to a state within a submachine and corresponds to equivalences in the product machine, that cannot be exploited to minimize the states of the submachine.

We can still utilize this don't care condition (or Boolean relation) to simplify the *logic* of Submachine 2. As described in Section 7.5.3, if two states s_1 and s_2 are equivalent in a given machine, we can exploit this condition at the logic level by specifying these states as next state don't cares (or Boolean relations) for all the edges $e \mid e.Next = s_1$ or $e.Next = s_2$. In Section 7.5.3, we did this uniformly for *all* the fanin edges of the two equivalent states. In our above example, we can specify a Boolean relation for the $n(EDGE1) = $ 01 *or* 10. Minimizing the logic of Submachine 2 under this Boolean relation can potentially result in simplification.

7.6.3   Invalid States and Edges

In the single machine case, we had invalid states which were not reachable from the starting states in the STG, and we could use these invalid states as don't cares to minimize the logic. In the interacting FSM case, we have invalid states and edges in the submachines, due to invalid states in the product machine.

Consider the state 11 in Submachine 2. It is an invalid state. This is because $xy, 11$ for all $x \in \{0, 1\}$ and $y \in \{0, 1\}$ are all invalid states in the product machine of Figure 7.17 (We have not shown these states in Figure 7.17, but it is clear that there are only five reachable states if the starting state is $00, 00$).

Consider the invalid state $10, 10$ in the product machine STG of Figure 7.17. This state can never be reached from the starting state. This implies that the edge from state 10 in Submachine 1 corresponding to the state input of 10 from Submachine 2 will never be traversed. This is an invalid edge in Submachine 1.

Invalid states and edges in submachines can be exploited in exactly the same manner as described in Section 7.5.2. For instance, we will specify $-10\ 10$ as a don't care when optimizing Submachine 2.

7.6.4   Searching for Unreachability and Compatibility

In this section, the algorithms used to identify unreachable states/untraversed edges and compatible states in mutually interacting machines are described in detail.

What gains can be accrued as a result of the fact that the FSMs communicate solely via their present states? The following lemma answers that question in part:

Lemma 7.6.1 *In an FSM network, if the only means of communication between the FSMs is all their state variables, assuming that are no don't cares on the primary inputs, the longest don't care input sequence that does not contain any other don't care sequences, for any machine, is of length two.*

Proof. Since the intermediate lines between FSMs are all the present state lines, a don't care sequence corresponds to sequences of states that don't occur. Invalid states correspond to don't care sequences of length one. Thus, don't care sequences of length two, that do not contain don't care sequences of length one, can only consist of valid states. Say, that s_1, s_2 and s_3 are states in a FSM, and that length-two sequences $s_1 \rightarrow s_2$ and $s_2 \rightarrow s_3$ are care sequences. Then the sequence $s_1 \rightarrow s_2 \rightarrow s_3$ has to be a care sequence. Hence, there are no don't care sequences of length three or more, that do not contain smaller don't care sequences. $\square$

The tangible benefit of this property is that the algorithm that identifies conditional compatibility between states only has to search for compatibility for every fanin edge traversed rather than for every fanin sequence traversed, as would be the case otherwise. Since the don't care sequences considered can at most be of length two, if two states are conditionally compatible under a fanin edge, e, it implies that there is a set of length-two don't care sequences such that e is the first vector in these don't care sequences. Further, *any* edge that is the first vector of a sequence differentiating the two states is the second vector of one of these length-two don't care sequences.

The procedures described below use the fact that given an edge in Submachine 1, one can easily find the corresponding edge in Submachine 2, and vice versa. For instance, an edge in Submachine 1 may be

traverse-edges(q_{mj}, q_{m1}, .. q_{mj-1}, q_{mj+1}, .. q_{mN}):
{
    foreach(unmarked fanout edge of q_{mj}, e_{mj}) {
        e_{mj} = $(i$ @$(q_{m1}, q_{m2}, .. q_{mj-1}, q_{mj+1}, .. q_{mN})$ @ $q_{mj})$;
        Mark e_{mj} ;
        foreach(mx = $m1$ to mN) {
            q'_{mx} = n$(i$ @$(q_{m1}, q_{m2}, .. q_{mx-1}, q_{mx+1}, .. q_{mN})$ @ $q_{mx})$;
        }
        traverse-edges(q'_{mj}, q'_{m1}, .. q'_{mj-1}, q'_{mj+1}, .. q'_{mN}) ;
    }
}

Figure 7.19: Procedure to traverse all reachable edges in a submachine

$(i$ @$(q_{m2}, q_{m3}, .. q_{mN})$ @ $q_{m1})^4$, where i is the input vector and q_{mk} is the state of Submachine mk. The corresponding edge in Submachine 2, for example, is then $(i$ @$(q_{m1}, q_{m3}, .. q_{mN})$ @ $q_{m2})$.

The procedure **traverse-edges**() of Figure 7.19 is a recursive procedure that traverses all the edges in a specified FSM given the starting state for each FSM. The first argument to **traverse-edges**() is the starting state of the FSM in which the edges are to be traversed. The other arguments are the set of starting states for the remaining FSMs. Every edge encountered during the enumeration is marked. The set of edges that remain unmarked when **traverse-edges**() is called with the actual starting state of each FSM as argument, is the set of untraversed edges for the specified FSM. Therefore, determining all the unreachable states/untraversed edges in all the FSMs requires calling **traverse-edges**() once for each FSM. In the pseudo-code shown below, **traverse-edges**() operates on the edges of Submachine mj. N is the total number of embedded FSMs. As in previous sections, n(e) denotes the next state that the edge e fans out to.

The procedure **trav-from-edge**() shown in Figure 7.20, finds

---

[4] The @ operator, as in $(i$ @ $q)$, represents a concatenation of the strings i and q. $(i$ @$(q_{m2}, q_{m3}, .. q_{mN})$ @ $q_{m1})$ is a concatenation of the strings i, q_{m2}, q_{m3} up to q_{mN} and the string q_{m1}.

trav-from-edge($e_{mj} = (i @(q_{m1}, \cdot\cdot q_{mj-1}, q_{mj+1}, \cdot\cdot q_{mN}) @ q_{mj})$):
{
    Unmark all edges ;
    Mark e_{mj} ;
    foreach($mx = m1$ to mN) {
        $q'_{mx} = $ n($i @(q_{m1}, q_{m2}, \cdot\cdot q_{mx-1}, q_{mx+1}, \cdot\cdot q_{mN}) @ q_{mx}$) ;
    }
    traverse-edges($q'_{mj}, q'_{m1}, \cdot\cdot q'_{mj-1}, q'_{mj+1}, \cdot\cdot q'_{mN}$) ;
}

Figure 7.20: Procedure for edge traversal from an edge

all the traversable edges, *given the traversal of some edge, e_{mj}*. **trav-from-edge**() will be used within a procedure that finds all conditionally compatible states.

Conditional compatibility is found using the procedure **find-all-compatible-states**() of Figure 7.21. G_{mj} denotes the STG of Submachine mj. Say that one is required to find all the compatibilities in Submachine mj. The basic idea is to traverse a certain fanin edge, say e_{mj} and identify all the edges in Submachine mj that cannot be traversed as a result. These untraversable edges are removed from G_{mj} to obtain G'_{mj}. By constructing G'_{mj}, one can use standard state minimization algorithms (*cf.* Section 7.2) to find conditional compatibilities. Note that one merely needs to find all states compatible to the fanin state of edge e_{mj} to find the fanin don't care set for edge e_{mj}. Unlike in state minimization, one does not require information as to the other compatibilities, nor does one have to merge maximal numbers of compatible states, as in Section 7.2.

Lemma 7.6.2 *The procedure* **find-all-compatible-states**() *finds all the conditionally and unconditionally compatible states for the embedded FSM specified as argument.*

Proof. According to Lemma 7.6.1 it is sufficient to search for compatibility between states only under the immediate fanin to the states. Also, it is apparent from the outline of the procedure **find-all-compatible-states**() given above that possibility of compatibility under all possible

find-all-compatible-states(G_{mj}, G_{m1}, .. G_{mj-1}, G_{mj+1},.. G_{mN}):
{
    foreach(valid edge $e_{mj} \in G_{mj}$) {
        $e_{mj} = (i \ @(q_{m1}, q_{m2}, .. q_{mj-1}, q_{mj+1}, .. q_{mN}) \ @ \ q_{mj})$;
        trav-from-edge(e_{mj}) ;
        $G'_{mj} \subset G_{mj}$ with only marked edges ;
        Find all states in G'_{mj} compatible to n(e_{mj}) ;
    }
}

Figure 7.21: Procedure to find unconditional and conditionally compatible states

immediate fanin is searched for. States that are unconditionally compatible can be thought of as being conditionally compatible under *all* their immediate fanin. Therefore, searching for all conditional compatibilities identifies all conditional as well as unconditional compatibilities. □

7.7 Conclusion

There are two basic advantages to a hierarchical or distributed representation of sequential circuits:

1. The representation of the total behavior in terms of multiple interacting STGs can be more compact than the STG of the overall lumped machine. For instance, in the case of using sum-of-products representations in State Transition Tables or Graphs, the state inputs to a Submachine **A**, from another Submachine **B**, need not be explicitly enumerated as minterms, since these lines are effectively primary inputs to **A**. Similarly, if the present state lines from **A** feed **B**, these are allowed to be don't cares in the cubes in the STG of **B**.

2. The number of states in each submachine is smaller than in the composite machine. Optimization procedures that operate on each

submachine separately can be faster.

The disadvantages are that in order to ensure global optimization capabilities in optimization procedures, more complicated analyses have to be carried out on the individual submachines, and the derivation of the don't cares may require large amounts of time. Further, the don't care sets may be very large, but large portions of the don't care set may not result in the simplification of logic. Directly deriving the *useful* portion of the don't care sets is a primary challenge in this area.

Chapter 8

Conclusions and Directions for Future Work

The focus of this book has been to describe the algorithms and Computer-Aided Design tools available for the synthesis of sequential circuits. As mentioned in the introduction, optimization is an integral facet of synthesis. We have concentrated on describing the theoretical and practical aspects of optimization-based sequential logic synthesis.

Testing and verification are two other facets of design that are strongly related to optimization-based synthesis. Circuit representations and data structures cut across all facets of design. At the combinational or sequential circuit level, Boolean functions are continually manipulated in various ways. The search for more efficient representations of Boolean functions is ceaseless, mainly because discovering such representations can have extensive impact on synthesis, test and verification problems. In Section 8.1, we will touch upon the use of Binary Decision Diagrams to represent Boolean functions.

The procedures described in this book operate at the symbolic level. The advantage of operating at the symbolic level is that global information regarding circuit functionality is readily available. The drawbacks are that accurately predicting circuit area or performance is difficult. Methods for sequential circuit optimization at the gate-level have been proposed. We describe some of these methods in Section 8.2.

In recent years, exploring the relationships between synthesis, verification and test has rapidly evolved into a rich and exciting area of research. Bartlett *et al* showed in [7] that using a *complete* don't care set could ensure 100% single stuck-at fault testability for combinational logic. Similarly, it has been shown that use of *complete* sequential don't care sets can ensure 100% non-scan single stuck-at fault testability [3, 25, 27, 28]. Ongoing work in this area is briefly summarized in Section 8.3.

Exploiting information available in a high-level circuit specification can potentially be of enormous help in managing logic-level complexity, especially for CPU-intensive optimization and test problems. For instance, one can envision efficient functional partitioning at the register-transfer level, that results in moderate-sized, possibly interacting, sequential or combinational circuits at the logic level, which can be independently optimized. Of course, a bad partitioning can result in an area-inefficient or low-performance design, if optimization is not carried out across the partitions. We touch upon methods that exploit register-transfer level in Section 8.4, including a sequential test generation and redundancy removal procedure. More detailed descriptions of verification and test algorithms relevant to sequential circuits can be found in a companion book [38].

We briefly describe available frameworks for sequential logic synthesis in Section 8.5.

8.1 Alternate Representations

While a large number of sequential logic synthesis programs available today use either a State Transition Graph (STG) representation or a two-level multiple-valued representation, it is clear that a number of sequential logic synthesis operations can also be performed while operating on a ordered Binary Decision Diagram (OBDD) [17] based characteristic-function representation [20]. For many of these operations, for example state reachability analysis, it is never necessary to resort to either a STG or two-level representation of the machine, *i.e.* it is never necessary to explicitly consider each individual state separately (sets of states are represented implicitly by Boolean equations). It is for such operations, that the OBDD-based characteristic-function representation clearly outperforms the STG representation.

The characteristic-function representation is attractive because it is practical even for controllers that contain some arithmetic circuitry, and is therefore applicable to a much larger class of circuits, than the classical STG representation where every state is explicitly stored. Unfortunately, the characteristic function makes it harder to gauge logic-level complexity as opposed to a sum-of-products representation of the STG. For instance, it is not clear how an optimum algorithm for finite state machine (FSM) decomposition could be developed, that targets the minimum number of product terms in a two-level realization, using characteristic functions alone. One of the important avenues for further research in sequential logic synthesis that can lead to a significant increase in the size of FSMs for which sequential optimization is practical is to investigate the extent to which the OBDD-based characteristic-function representation can be used to address synthesis problems. At any level of abstraction, the ability to simulate at that level is a key necessary condition for enabling synthesis at that level. The fact the OBDD-based characteristic-function representations are viable for traversing FSMs and conducting reachability analysis was illustrated in [20]. The application of characteristic-function representations for detecting equivalent-state pairs was illustrated in [63]. It remains to be seen whether the characteristic-function representation can be used effectively for FSM decomposition and state encoding.

8.2   Optimization at the Logic Level

We have alluded several times in this book to the two major problems that plague the synthesis from symbolic descriptions of sequential circuits. The first is the one described in the previous section. The second is that the cost function is dependent on the particular synthesis methodology that will be used at the combinational level after encoding. To get around both these problems, approaches that use retiming and resynthesis to optimize the sequential circuit at the logic level [66, 72], have been proposed. Unfortunately, the techniques proposed so far have not been very effective in being able to realize substantial optimization on controller-type sequential circuits. These techniques do hold promise and are a subject of ongoing research. A possible alternative approach to sequential synthesis at the logic level is the following. The circuit to be optimized can be considered as a set of interacting sequential circuits, partitioned so that the logic to be optimized lies in one of the subcircuits. The circuits are assumed to interact through their present states. Strong or Boolean division [14, 31] can now be used to *re-encode a set of wires* that feed the subcircuit to be optimized at the same time. The decoding logic block can then be appropriately substituted into the necessary subcircuits to maintain functionality. While this is guaranteed to improve the performance of the subcircuits from which the logic was extracted, a drawback is that the effect of the resubstitution is not very deterministic.

The procedures presented in Chapters 3 through Chapter 6 and most other current sequential synthesis procedures require some form of State Transition Graph representation to operate from. Recognizing this, procedures for efficient State Transition Graph extraction from logic-level descriptions have been developed [5], and have described in a companion book [38]. Using these procedures, it has been possible to increase the size of the sequential circuits for which current sequential logic synthesis strategies are viable.

8.3   Don't Cares and Testability

The connection between don't cares and single stuck-at fault redundancy in combinational circuits was explored in [7, 9]. The corresponding relationships between sequential don't cares and non-scan single stuck-at fault testability has been a subject of recent investigation (*e.g.* [3, 27]).

Devadas *et al* in [27] defined a complete don't care set for a FSM represented by a single STG, which if optimally exploited results in full non-scan testability. The procedure described requires repeated logic minimization, and can be quite CPU-intensive. This procedure was generalized to be applicable at the register-transfer level in [37]. As described in the next section, this generalization significantly improved the efficiency of the procedure.

If an implementation was realized as a network of interacting sequential circuits (possibly using the algorithms presented in Chapter 6), it may be required to maintain the original structure for various reasons while applying the synthesis-for-test procedures. Procedures that exploit sequential don't cares in interacting sequential circuits were presented in Chapter 7. It was shown by Ashar, Devadas and Newton in [3] that exploiting sequential don't cares is necessary for the synthesis of non-scan irredundant interacting sequential circuits. A complete sequential don't care set can ensure complete testability. By virtue of these procedures, it is possible to generate don't cares for each component submachine based on its interaction with the environment, and subsequently to optimize each submachine under its don't care set, thereby maintaining the overall structure. However, the efficiency of these logic-level procedures for large sequential circuits is questionable.

Computing sequential don't cares is just one half of the synthesis for test problem. The other half has to do with exploiting them optimally during combinational optimization. Image computation methods, originally developed for FSM verification [20], have recently been used for computing local don't cares in multilevel combinational circuits [86]. These techniques have made it possible to use external don't cares effectively for multilevel circuits without having to first collapse the circuit, thereby making the exploitation of sequential don't cares practical.

The generalization of these procedures to operate at higher levels of abstraction to improve efficiency is a subject for future research.

8.4   Exploiting Register-Transfer Level Information

In recent work by Ghosh, Devadas and Newton [37], it was shown that information at the register-transfer-level could be used to simplify the test generation and synthesis-for-test problems for sequential circuits. Justifying particular sets of values at the outputs of arithmetic circuits can be done much more efficiently given information as to the functionality of the arithmetic unit. The efficiency of state reachability analysis can also be dramatically improved. The work of [37] has been described in detail in a companion book [38].

Functional information is useful for other applications as well. For instance, efficient orderings for ordered Binary Decision Diagrams can be found by analyzing register-transfer-level descriptions of the circuit [18]. Efficient orderings cannot always be found by algorithms that operate purely at the logic level. We believe that increasing use of higher level information in synthesis, verification and test problems will be necessary to manage the increasing complexity of VLSI circuits.

8.5   Sequential Logic Synthesis Systems

Sequential logic synthesis, test and verification are currently active areas of research both in academia and industry. Projects in several research institutions and laboratories involve the integration of CAD tools that operate on sequential circuits, at various levels of abstraction, into cohesive systems. The system SIS being developed at U. C. Berkeley, contains the full functionality of MIS-II and includes algorithms for logic-level retiming [59], peripheral retiming [66], state encoding [26, 94], and sequential logic verification [20, 91].

A synthesis system called FLAMES has been developed jointly at MIT and U. C. Berkeley which incorporates most of the algorithms described in this book, and as well as some of the testing and verification algorithms described in a companion book [38]. From the start, FLAMES was meant to be a sequential synthesis system that would be able to manipulate networks of interacting sequential circuits. The organization and the main features of FLAMES have been described in [2].

In the typical synthesis/re-synthesis scenario, a sequential circuit or a set of interacting sequential circuits is read into FLAMES for optimization. The logic that needs to be optimized is identified by first identifying, for example, the critical paths or possibly the difficult-to-test faults in the circuit. The sequential circuit is then partitioned accordingly and the symbolic representation for the subcircuits to be optimized is extracted. Various sequential and combinational optimization strategies are then applied to achieve the desired specifications. Finally, the optimized circuit is verified against the original circuit for equivalence.

Bibliography

[1] D. B. Armstrong. A Programmed Algorithm for Assigning Internal Codes to Sequential Machines. In *IRE Transactions on Electronic Computers*, volume EC-11, pages 466–472, August 1962.

[2] P. Ashar. *Synthesis of Sequential Circuits for VLSI Design*. PhD thesis, University of California, Berkeley, August 1991.

[3] P. Ashar, S. Devadas, and A. R. Newton. Irredundant Interacting Sequential Machines Via Optimal Logic Synthesis. In *IEEE Transactions on Computer-Aided Design of Integrated Circuits and Systems*, volume 10, pages 311–325, March 1991.

[4] P. Ashar, S. Devadas, and A. R. Newton. Optimum and Heuristic Algorithms for a Problem of Finite State Machine Decomposition. In *IEEE Transactions on Computer-Aided Design of Integrated Circuits and Systems*, volume 10, pages 296–310, March 1991.

[5] P. Ashar, A. Ghosh, S. Devadas, and A. R. Newton. Implicit State Transition Graphs: Applications to logic synthesis and test. In *Proceedings of the International Conference on Computer Aided Design*, pages 84–87, November 1990.

[6] R. L. Ashenhurst. The Decomposition of Switching Functions. In *Proceedings International Symposium on Theory of Switching Functions*, pages 74–116, April 1959.

[7] K. Bartlett, R. K. Brayton, G. D. Hachtel, R. M. Jacoby, C. R. Morrison, R. L. Rudell, A. Sangiovanni-Vincentelli, and A. R. Wang. Multi-level Logic Minimization using Implicit Don't Cares. In *IEEE Transactions on Computer-Aided Design of Integrated Circuits and Systems*, volume 7, pages 723–740, June 1988.

[8] D. Bostick, G. Hachtel, R. Jacoby, M. Lightner, P. Moceyunas,
    C. Morrison, and D. Ravenscroft. The Boulder Optimal Logic
    Design System. In *Proceedings of the International Conference on
    Computer Aided Design*, pages 62–65, November 1987.

[9] D. Brand. Redundancy and Don't Cares in Logic Synthesis. In
    IEEE Transactions on Computers, volume C-32, pages 947–952,
    October 1983.

[10] R. Brayton, R. Rudell, A. Sangiovanni-Vincentelli, and A. Wang.
     MIS: A Multiple-level Logic Optimization System. In *IEEE Trans-
     actions on Computer-Aided Design of Integrated Circuits and Sys-
     tems*, volume CAD-6, pages 1062–1081, November 1987.

[11] R. Brayton, R. Rudell, A. Sangiovanni-Vincentelli, and A. Wang.
     Multi-Level Logic Optimization and The Rectangular Covering
     Problem. In *Proceedings of the International Conference on Com-
     puter Aided Design*, pages 66–69, November 1987.

[12] R. K. Brayton, J. D. Cohen, G. D. Hachtel, B. M. Trager, and
     D.Y. Yun. Fast recursive Boolean function manipulation. In *Pro-
     ceedings of the 1982 International Symposium on Circuits and Sys-
     tems*, pages 58–62, April 1982.

[13] R. K. Brayton, G. D. Hachtel, C. McMullen, and A. Sangiovanni-
     Vincentelli. *Logic Minimization Algorithms for VLSI Synthesis*.
     Kluwer Academic Publishers, Boston, Massachusetts, 1984.

[14] R. K. Brayton, G. D. Hachtel, and A. L. Sangiovanni-Vincentelli.
     Multilevel Logic Synthesis. In *Proceedings of the IEEE*, pages 264–
     300, February 1990.

[15] R. K. Brayton and C. McMullen. The decomposition and facoriza-
     tion of Boolean expressions. In *Proceedings of the International
     Symposium on Circuits and Systems*, pages 49–54, Rome, May
     1982.

[16] R. K. Brayton and F. Somenzi. Boolean Relations and the In-
     complete Specification of Logic Networks. In *IFIP International
     Conference on Very Large Scale Integration*, pages 231–240, Au-
     gust 1989.

[17] R. Bryant. Graph-based algorithms for Boolean function manipulation. In *IEEE Transactions on Computers*, volume C-35, pages 677–691, August 1986.

[18] H. Cho, G. Hachtel, S-W. Jeong, B. Plessier, E. Schwarz, and F. Somenzi. ATPG Aspects of FSM Verification. In *Proceedings of the International Conference on Computer Aided Design*, pages 134–137, November 1990.

[19] H. C.Lai and S. Muroga. Automated Logic Design of MOS Networks. In *Advances in Information Systems Science*, volume 9, pages 287–335. Plenum Press, New York - London, 1970.

[20] O. Coudert, C. Berthet, and J. C. Madre. Verification of Sequential Machines using Functional Boolean Vectors. In *IMEC-IFIP International Workshop on Applied Formal Methods for Correct VLSI Design*, pages 111–128, November 1989.

[21] M. Dagenais, V. K. Agarwal, and N. Rumin. McBOOLE: A Procedure for Exact Boolean Minimization. In *IEEE Transactions on Computer-Aided Design of Integrated Circuits and Systems*, volume CAD-5, pages 229–237, January 1986.

[22] J. Darringer, D. Brand, J. Gerbi, W. Joyner, and L. Trevillyan. LSS: A System for Production Logic Synthesis. In IBM *J. Res. Develop.*, volume 28, pages 537–545, September 1984.

[23] E. S. Davidson. An algorithm for NAND Decomposition Under Network Constraints. In *IEEE Transactions on Computers*, volume C-18, pages 1098–1109, December 1969.

[24] S. Devadas. Approaches to Multi-Level Sequential Logic Synthesis. In *Proceedings of the 26th Design Automation Conference*, pages 270–276, June 1989.

[25] S. Devadas and K. Keutzer. A Unified Approach to the Synthesis of Fully Testable Sequential Machines. In *IEEE Transactions on Computer-Aided Design of Integrated Circuits and Systems*, volume 10, pages 39–50, January 1991.

[26] S. Devadas, H-K. T. Ma, A. R. Newton, and A. Sangiovanni-Vincentelli. MUSTANG: State assignment of finite state machines targeting multi-level logic implementations. In *IEEE Transactions on Computer-Aided Design of Integrated Circuits and Systems*, volume 7, pages 1290–1300, December 1988.

[27] S. Devadas, H-K. T. Ma, A. R. Newton, and A. Sangiovanni-Vincentelli. Irredundant Sequential Machines Via Optimal Logic Synthesis. In *IEEE Transactions on Computer-Aided Design of Integrated Circuits and Systems*, volume 9, pages 8–18, January 1990.

[28] S. Devadas, H-K. Tony Ma, and A. R. Newton. Redundancies and Don't Cares in Sequential Logic Synthesis. In *Journal of Electronic Testing: Theory and Applications*, volume 1, pages 15–30, February 1990.

[29] S. Devadas and A. R. Newton. Decomposition and Factorization of Sequential Finite State Machines. In *IEEE Transactions on Computer-Aided Design of Integrated Circuits and Systems*, volume 8, pages 1206–1217, November 1989.

[30] S. Devadas and A. R. Newton. Exact Algorithms for Output Encoding, State Assignment and Four-Level Boolean Minimization. In *IEEE Transactions on Computer-Aided Design of Integrated Circuits and Systems*, volume 10, pages 13–27, January 1991.

[31] S. Devadas, A. R. Wang, A. R. Newton, and A. Sangiovanni-Vincentelli. Boolean Decomposition in Multi-Level Logic Optimization. In *Journal of Solid State Circuits*, volume 24, pages 399–408, April 1989.

[32] T. A. Dolotta and E. J. McCluskey. The Coding of Internal States of Sequential Machines. In *IEEE Transactions on Electronic Computers*, volume EC-13, pages 549–562, October 1964.

[33] X. Du, G. Hachtel, B. Lin, and A. R. Newton. MUSE: A MUltilevel Symbolic Encoding Algorithm for State Assignment. In *IEEE Transactions on Computer-Aided Design of Integrated Circuits and Systems*, volume 10, pages 28–38, January 1991.

[34] M. Foster. Partitioning real finite automata. In *AT&T Bell Laboratories Internal Memorandum*, August 1988.

[35] M.R. Garey and D.S. Johnson. *Computers and Intractability, A Guide to the Theory of NP-Completeness*. W.H. Freeman, New York, second edition, 1979.

[36] A. Ghosh, S. Devadas, and A. R. Newton. Heuristic Minimization of Boolean Relations Using Testing Techniques. In *Proceedings of the Int'l Conference on Computer Design*, pages 277–281, October 1990.

[37] A. Ghosh, S. Devadas, and A. R. Newton. Sequential Test Generation at the Register Transfer and Logic Levels. In *Proceedings of the 27th Design Automation Conference*, pages 580–586, June 1990.

[38] A. Ghosh, S. Devadas, and A. R. Newton. *Sequential Logic Testing and Verification*. Kluwer Academic Publishers, Boston, Massachusetts, 1991.

[39] S. Ginsburg. A Synthesis Technique for Minimal State Sequential Machines. In *IRE Transactions on Electronic Computers*, volume EC-8, pages 13–24, March 1959.

[40] D. Gregory, K. Bartlett, A. de Geus, and G. Hachtel. SOCRATES: A System for Automatically Synthesizing and Optimizing Combinational Logic. In *Proceedings of the 23rd Design Automation Conference*, pages 79–85, June 1986.

[41] J. Hartmanis. Symbolic Analysis of a Decomposition of Information Processing. In *Information Control*, volume 3, pages 154–178, June 1960.

[42] J. Hartmanis. On the State Assignment Problem for Sequential Machines I. In *IRE Transactions on Electronic Computers*, volume EC-10, pages 157–165, June 1961.

[43] J. Hartmanis and R. E. Stearns. Some Dangers in the State Reduction of Sequential Machines. In *Information and Control*, volume 5, pages 252–260, September 1962.

[44] J. Hartmanis and R. E. Stearns. *Algebraic Structure Theory of Sequential Machines.* Prentice-Hall, Englewood Cliffs, New Jersey, 1966.

[45] F. C. Hennie. *Finite-State Models for Logical Machines.* Wiley, New York, New York, 1968.

[46] S. J. Hong, R. G. Cain, and D. L. Ostapko. MINI: A Heuristic Approach for Logic Minimization. In *IBM Journal of Research and Development*, volume 18, pages 443–458, September 1974.

[47] J. Hopcroft. A $n \log n$ algorithm for minimizing states in a finite automaton. In *Theory of Machines and Computations, Z. Kohavi and A. Paz (eds.)*, pages 189–196. Academic Press, New York, 1971.

[48] D. A. Huffman. The Synthesis of Sequential Switching Circuits. In *J. Franklin Institute*, volume 257, no. 3, pages 161–190, 1954.

[49] D. A. Huffman. The Synthesis of Sequential Switching Circuits. In *J. Franklin Institute*, volume 257, no. 4, pages 275–303, 1954.

[50] R. M. Karp. Some Techniques for the State Assignment of Synchronous Sequential Machines. In *IEEE Transactions on Electronic Computers*, volume EC-13, pages 507–518, October 1964.

[51] J. Kim and M. M. Newborn. The Simplification of Sequential Machines with Input Restrictions. In *IEEE Transactions on Computers*, volume C-20, pages 1440–1443, December 1972.

[52] S. C. Kleene. Representation of Events in Nerve Nets and Finite Automata. In *Automata Studies*, pages 3–42. Princeton University Press, Princeton, New Jersey, 1956.

[53] Z. Kohavi. Secondary State Assignment for Sequential Machines. In *IEEE Transactions on Electronic Computers*, volume EC-13, pages 193–203, June 1964.

[54] Z. Kohavi. *Switching and Finite Automata Theory.* McGraw-Hill Book Company, New York, New York, second edition, 1978.

[55] K. Krohn and J. Rhodes. Algebraic Theory of Machines. In *Proceedings Symposium on Mathematical Theory of Automata*. Polytechnic Press, N.Y., 1962.

[56] L. Lavagno, S. Malik, R. Brayton, and A. Sangiovanni-Vincentelli. MIS-MV: Optimization of Multi-level Logic with Multiple-valued Inputs. In *Proceedings of the International Conference on Computer Aided Design*, pages 560–563, November 1990.

[57] L. Lavagno, S. Malik, R. Brayton, and A. Sangiovanni-Vincentelli. Optimization of multi-level logic with multiple-valued inputs. In *Technical report, Electronics Research Laboratory, U. C. Berkeley*, August 1990.

[58] E. L. Lawler. An Approach to Multilevel Boolean Minimization. In *Journal of the Association for Computing Machinery*, volume 11, pages 283–295, July 1964.

[59] C. E. Leiserson, F. M. Rose, and J. B. Saxe. Optimizing Synchronous Circuitry by Retiming. In *Proceedings of 3rd CalTech Conference on VLSI*, pages 23–36, March 1983.

[60] B. Lin and A. R. Newton. A Generalized Approach to the Constrained Cubical Embedding Problem. In *Proceedings of the Int'l Conference on Computer Design: VLSI in Computers and Processors*, pages 400–403, October 1989.

[61] B. Lin and A. R. Newton. Synthesis of Multiple-Level Logic From Symbolic High-Level Description Languages. In *IFIP International Conference on Very Large Scale Integration*, pages 187–196, August 1989.

[62] B. Lin and F. Somenzi. Minimization of Symbolic Relations. In *Proceedings of the International Conference on Computer Aided Design*, pages 88–91, November 1990.

[63] B. Lin, H. Touati, and A. R. Newton. Don't Care Minimization of Multi-level Sequential Networks. In *Proceedings of the International Conference on Computer Aided Design*, pages 414–417, November 1990.

[64] R. Lisanke, editor. *FSM Benchmark Suite*. Microelectronics Center of North Carolina, Research Triangle Park, North Carolina, 1987.

[65] S. Malik, R. Brayton, and A. Sangiovanni-Vincentelli. Encoding Symbolic Inputs for Multi-level Logic Implementation. In *IFIP International Conference on Very Large Scale Integration*, pages 221–230, August 1989.

[66] S. Malik, E. M. Sentovich, R. Brayton, and A. Sangiovanni-Vincentelli. Retiming and Resynthesis: Optimizing Sequential Circuits Using Combinational Techniques. In *IEEE Transactions on Computer-Aided Design of Integrated Circuits and Systems*, volume 10, pages 74–84, January 1991.

[67] M. P. Marcus. Deriving Maximal Compatibles using Boolean Algebra. In *IBM Journal of Research and Development*, volume 8, pages 537–538, November 1964.

[68] E. J. McCluskey. Minimization of Boolean functions. In *Bell Lab. Technical Journal*, volume 35, pages 1417–1444. Bell Lab., November 1956.

[69] W. S. McCulloch and W. Pitts. A Logical Calculus of the Ideas Immanent in Nervous Activity. In *Bull. Math. Biophysics*, volume 5, pages 115–133, 1943.

[70] G. H. Mealy. A Method of Synthesizing Sequential Circuits. In *Bell System Tech. Journal*, volume 34, pages 1045–1079, September 1955.

[71] G. De Micheli. Symbolic Design of Combinational and Sequential Logic Circuits implemented by Two-Level Macros. In *IEEE Transactions on Computer-Aided Design of Integrated Circuits and Systems*, volume CAD-5, pages 597–616, September 1986.

[72] G. De Micheli. Synchronous Logic Synthesis: Algorithms for Cycle-Time Minimization. In *IEEE Transactions on Computer-Aided Design of Integrated Circuits and Systems*, volume 10, pages 63–73, January 1991.

[73] G. De Micheli, R. K. Brayton, and A. Sangiovanni-Vincentelli. Optimal State Assignment of Finite State Machines. In *IEEE Transactions on Computer-Aided Design of Integrated Circuits and Systems*, volume CAD-4, pages 269–285, July 1985.

[74] G. De Micheli, A. Sangiovanni-Vincentelli, and T. Villa. Computer-Aided Synthesis of PLA-based Finite State Machines. In *Proceedings of the International Conference on Computer Aided Design*, pages 154–156, November 1983.

[75] E. F. Moore. Gedanken-experiments on sequential machines. In *Automata Studies*, pages 129–153, Princeton, New Jersey, 1956. Princeton University Press.

[76] M. C. Paull and S. H. Unger. Minimizing the number of states in Incompletely Specified Sequential Circuits. In *IRE Transactions on Electronic Computers*, volume EC-8, pages 356–357, September 1959.

[77] W. Quine. The problem of simplifying truth functions. In *Amer. Math. Monthly*, volume 59, pages 521–531, 1952.

[78] W. V. Quine. A way to simplify truth functions. In *Am. Math. monthly*, volume 62, pages 627–631, Nov. 1955.

[79] P. J. Roth and R. M. Karp. Minimization Over Boolean Graphs. In *IBM Journal of Research and Development*, volume 6, pages 227–238, April 1962.

[80] R. Rudell and A. Sangiovanni-Vincentelli. Exact Minimization of Mutiple-Valued Functions for PLA Optimization. In *Proceedings of the International Conference on Computer Aided Design*, pages 352–355, November 1986.

[81] R. Rudell and A. Sangiovanni-Vincentelli. Multiple-valued Minimization for PLA Optimization. In *IEEE Transactions on Computer-Aided Design of Integrated Circuits and Systems*, volume CAD-6, pages 727–751, September 1987.

[82] A. Saldanha and R. H. Katz. PLA Optimization Using Output Encoding. In *Proceedings of the International Conference on Computer Aided Design*, pages 478–481, November 1988.

[83] A. Saldanha, T. Villa, R. K. Brayton, and A. Sangiovanni-Vincentelli. A Framework for Satisfying Input and Output Constraints. In *Proceedings of the 28th Design Automation Conference*, pages 170–175, June 1991.

[84] A. Saldanha, A. Wang, R. Brayton, and A. Sangiovanni-Vincentelli. Multi-Level Logic Simplification using Don't Cares and Filters. In *Proceedings of the 26th Design Automation Conference*, pages 277–282, June 1989.

[85] G. Saucier, M. C. Depaulet, and P. Sicard. ASYL: A rule-based system for controller synthesis. In *IEEE Transactions on CAD*, volume CAD-6, pages 1088–1097, November 1987.

[86] H. Savoj, R. K. Brayton, and H. J. Touati. Use of Image Computation Techniques in Extracting Local Don't Cares and Network Optimization. In *International Workshop on Logic Synthesis*, May 1991.

[87] C. E. Shannon. The synthesis of two-terminal switching circuits. *Bell Lab. Technical Journal*, pages 59–98, 1949.

[88] F. Somenzi and R.K. Brayton. An Exact Minimizer for Boolean Relations. In *Proceedings of the International Conference on Computer Aided Design*, pages 316–319, November 1989.

[89] R. E. Stearns and J. Hartmanis. On the State Assignment Problem for Sequential Machines II. In *IRE Transactions on Electronic Computers*, volume EC-10, pages 593–604, December 1961.

[90] P. Tison. Generalization of Consensus Theory and Application to the minimization of Boolean Functions. In *IEEE Transactions on Computers*, volume 16, pages 446–450, August 1967.

[91] H. Touati, H. Savoj, B. Lin R. K. Brayton, and A. Sangiovanni-Vincentelli. Implicit State Enumeration of Finite State Machines using BDDs. In *Proceedings of the International Conference on Computer Aided Design*, pages 130–133, November 1990.

[92] J. H. Tracey. Internal State Assignment for Asynchronous Sequential Machines. In *IEEE Transactions on Electronic Computers*, volume EC-15, pages 551–560, August 1966.

[93] S. H. Unger. *Asynchronous Sequential Switching Theory.* Wiley, New York, New York, 1969.

[94] T. Villa and A. Sangiovanni-Vincentelli. NOVA: State Assignment of Finite State Machines for Optimal Two-Level Logic Implementations. In *IEEE Transactions on Computer-Aided Design of Integrated Circuits and Systems*, volume 9, pages 905–924, September 1990.

[95] W. Wolf, K. Keutzer, and J. Akella. A Kernel Finding State Assignment Algorithm for Multi-Level Logic. In *Proceedings of the 25th Design Automation Conference*, pages 433–438, June 1988.

[96] W. Wolf, K. Keutzer, and J. Akella. Addendum to A Kernel Finding State Assignment Algorithm for Multi-Level Logic. In *IEEE Transactions on CAD*, volume 8, pages 925–927, August 1989.

[97] S. Yang and M. Ciesielski. On the Relationship between Input Encoding and Logic Minimization. In *Proceedings of the 23rd Hawaii International Conference on the System Sciences*, pages 377–386, January 1990.

[98] M. Yoeli. The Cascade Decomposition of Sequential Machines. In *IRE Transactions Electronic Computers*, volume EC-10, pages 587–592, April 1961.

[99] M. Yoeli. Cascade Parallel Decomposition of Sequential Machines. In *IRE Transactions on Electronic Computers*, volume EC-12, pages 322–324, June 1963.

[100] H. P. Zeiger. *Loop-free Synthesis of Finite-State Machines.* PhD thesis, Massachusetts Institute of Technology, Cambridge, 1964.

Index